WITHD

THE MILITARY-INDUSTRIAL COMPLEX: A Reassessment

SAGE RESEARCH PROGRESS SERIES ON
WAR, REVOLUTION, AND PEACEKEEPING

Charles C. Moskos, Jr., SERIES EDITOR

Sponsored by the Inter-University Seminar on Armed Forces and Society

Chairman: MORRIS JANOWITZ
Associate Chairman: CHARLES C. MOSKOS, Jr.
Executive-Secretary: SAM C. SARKESIAN
Executive Committee:
    Albert D. Biderman   Peter Karsten
    Henry Bienen   Catherine Kelleher
    Martin Blumenson   Laurence Radway
    Alexander George   T. Alden Williams
Counsel: Maury Jacobs
Membership Committee:
    David Eaton
    Thomas Blau
Treasurer: James Linger

THE MILITARY-INDUSTRIAL COMPLEX

A Reassessment

Edited by SAM C. SARKESIAN

*This volume was produced in cooperation with
The Center for Policy Study at The University of Chicago*

SAGE RESEARCH PROGRESS SERIES ON
WAR, REVOLUTION, AND PEACEKEEPING

Volume II (1972)

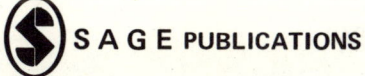 S A G E PUBLICATIONS   Beverly Hills / London

ACKNOWLEDGMENTS

The Center for Policy Study at The University of Chicago provides a forum for the review and public discussion of major foreign and domestic policy issues. The Center's Arms Control and Foreign Policy Seminar, funded by the Ford Foundation, sponsored an Institute on "The Military-Industrial Complex and Arms Control," December 3-4, 1971. The papers by Charles Moskos, Jr., Charles Wolf, Jr., Albert D. Biderman, Morris Janowitz, Seymour Melman, Bruce M. Russett, K. Wayne Smith and J. N. Merritt, and Adam Yarmolinsky were originally prepared for this Institute. They are printed in this volume by permission of the Center for Policy Study.

The Inter-University Seminar on Armed Forces and Society is a group of scholars from the academic, military, and research fields whose interests include the study of subjects relevant to armed forces and society. Its main purpose is to provide a focal point for the exchange of ideas and viewpoints not only on the internal structure of the military establishment and its relationship with society, but also on a comparative approach to the military in both industrial and developing nations. The Inter-University Seminar is supported by the Russell Sage Foundation. The paper by Lawrence Korb was prepared for the Sixth Annual Conference of the Inter-University Seminar and is printed here by permission.

The paper by Stanley Lieberson appeared in *The American Journal of Sociology*, Volume 76, Number 4 (January 1971) and is reprinted here by permission of The University of Chicago Press.

Copyright © 1972 by Sage Publications, Inc.

All rights reserved. No part of this book may be reproduced or utilized in any form or by any means, electronic or mechanical, including photocopying, recording, or by any information storage and retrieval system, without permission in writing from the publisher.

For information address:

SAGE PUBLICATIONS, INC.
275 South Beverly Drive
Beverly Hills, California 90212

SAGE PUBLICATIONS LTD
St George's House / 44 Hatton Garden
London E C 1

Printed in the United States of America

International Standard Book Number 0-8039-0134-8(C); 0-8039-0135-6(P)

Library of Congress Catalog Card No. 78-183960

FIRST PRINTING

CONTENTS

Sam C. Sarkesian   vi

INTRODUCTION

I. CONCEPTS AND REALITIES

Charles C. Moskos, Jr.   3

THE MILITARY-INDUSTRIAL COMPLEX: THEORETICAL ANTECEDENTS AND CONCEPTUAL CONTRADICTIONS

Charles Wolf, Jr.   25

MILITARY-INDUSTRIAL SIMPLICITIES, COMPLEXITIES AND REALITIES

Stanley Lieberson   53

AN EMPIRICAL STUDY OF MILITARY-INDUSTRIAL LINKAGES

Albert D. Biderman   95

RETIRED SOLDIERS WITHIN AND WITHOUT THE MILITARY-INDUSTRIAL COMPLEX

II. STRATEGIC ALTERNATIVES

Morris Janowitz   127

STRATEGIC DIMENSIONS OF AN ALL VOLUNTEER ARMED FORCE

Seymour Melman   167

ALTERNATIVE STRATEGIES AND BUDGETS FOR MILITARY SECURITY

Bruce M. Russett   201

A COUNTERCOMBATANT DETERRENT? FEASIBILITY, MORALITY, AND ARMS CONTROL

III. ARMS AND CONFLICT CONTROL

K. Wayne Smith and J. N. Merritt   245

SOME COMPLEXITIES OF ARMS CONTROL PLANNING

IV. DECISION-MAKING AND POLITICAL LEADERSHIP

Adam Yarmolinsky   277

THE PRESIDENT, THE CONGRESS, AND ARMS CONTROL

Lawrence J. Korb   301

THE SECRETARY OF DEFENSE AND THE JOINT CHIEFS OF STAFF: THE BUDGETARY PROCESS

INTRODUCTION

Sam C. Sarkesian

Although the study of elites has been a standard part of political and sociological discourse, it has been only since the Korean War period and the Eisenhower Presidency that the concept of military industrial complex wedded to an elite collusion theory has gained intellectual prominence. Ranging from polemical cliches to serious scholarly effort, the literature on this subject provides few coherent directions other than the fact that there exists some type of convergence of interests between industrial corporations and those either in the military hierarchy or those associated with it. The military-industrial complex however, does more than suggest a convergence of interests. Indeed, the ramifications of this idea enter into a diverse range of areas from strategic dimensions to domestic economic priorities, which in themselves are the subject of debate and disagreement.

Fundamentally however, the germinal issue is the nature and scope of the total U.S. defense effort. Some observers and critics presume the defense establishment has become so pervasive in the political system and in the international field that it has distorted the representative and democratic character of our political system while increasing international tensions. In this view, many would suggest that the defense establishment has generated a series of forces reinforcing its growth, while fostering the expansion of defense industrial production, and the perpetuation of those in power. Some of the most persistent critics of the Department of Defense point to its life and death control over many industries within the country and its tremendous propaganda apparatus that pereptuates the image of the necessity of an armed America.

> More awesome than his control over the actions of the military force is the power he [Secretary of Defense] holds over the spending of a budget of more than $50 billion a year. This is a power that can and does touch nearly every facet of our society, including the businesses and political community. It is a power that we have been warned should never be permitted to fall into the hands of any authoritarian-minded military man, and there are those who question the wisdom of concentrating such powers in the hands of even the best-motivated civilian Secretary of Defense.*

Thus, the issue of a military-industrial complex is more encompassing than the nature of its corporateness or the character of its elite. The issues which are raised involve not only the total defense effort, but costs in terms of social welfare and domestic economies. A number of economists, sociologists as well as political scientists, for example, attack the military-industrial complex on the grounds that it distorts the functioning of the political and social systems by providing resources and power to a corporate structure that is unrepresentative and virtually immune from the checks and balances of representative government. Furthermore, it is charged that the existence of the military-industrial complex not only is based on relative economic inefficiency, but draws resources that do not produce final products that have an economically useful life within the domestic economy. By drawing scarce resources, domestic economic priorities go unfulfilled, stimulating social injustice and inequality—resulting in the social injustice perpetuated by corporate control of political America. In the final analysis, a number of observers charge that the military-industrial complex, whatever its composition and whatever its corporate structure, has become a dysfunctional element within the political system because of its rapid growth to a position of dominance in decision making and in its selfish motivation for profit, at the expense of the public welfare.

From these observations, a wide range of ideological positions can be offered to explain the existence and growth of the military-industrial complex, as well as providing a basis for attacking it. Similarly, a variety of conceptual approaches can be formulated on the same premises. Regardless of the ideological position or academic orientation, however, most observers would accept the thesis that careful analyses and critique is necessary regarding the inter-relationships between the defense establishment, the domestic economy, and the decision making processes because there has been a lack of empirical research. Most would also agree that there is overlap, if not convergence, of interests between the military hierarchy, some industrial corporations, and vested interests within the government. There is however, much disagreement concerning the composition and extension of these interests, the resulting impact, and means to control, manage, and reduce the military-industrial complex.

It is much easier to identify the general nature of the problem than it is to provide realistic analysis. The characteristics of the military-industrial complex are such that the distinguishable components are difficult to untangle. Although a scholarly assessment is difficult, there is utility in analyzing the military-industrial complex from at least two levels. The first, a problem approach, is one that focuses on the study of specific issues, or various components, within the military establishment in order to identify their utility and impact on the defense effort. This approach should culminate in a number of objective issue oriented studies identifying and clarifying the boundaries and interrelationships of military and industrial oriented interests. The second, a sub-system approach, views the defense establishment as a sub-system within the total political system, with its own rules of the game, values, and interests. Such an approach is concerned with the identification of the decision making process within the sub-system, its sources of power and control over other sub-systems such as the Congressional establishment or the federal bureaucratic structure. It is, in addition, concerned with the degree of control of the military over the total political system. This approach also compares the sub-systems and their relationships to each other and to the political system, thereby identifying the dysfunctional nature of any sub-system, i.e., the defense establishment. The papers in this volume are examples of both approaches.

This volume seeks to reassess the military-industrial complex, not only with respect to conceptual clarity, but to the empirical and substantive issues that are at the core of the debate. In addition to a number of issues dealing with military-industrial linkages, it seeks to explore the potential role of the military in alternative strategies and arms limitations. The authors of the papers in this volume do not postulate a unified view of the military-industrial complex. Nor do they agree on the impact of the defense effort on domestic society. Nevertheless, the papers not only present a fundamental critique of military-industrial interpenetrations; they also agree that hard data are required to understand the components that have important linkages of industry with the defense establishment and its range of activities both domestic and international.

The majority of the contributions to this volume were originally presented at a conference on "The Military-Industrial Complex and Arms Control" sponsored by the Center for Policy Study, Arms Control and Foreign Policy Seminar, the University of Chicago, held on December 3-5, 1971. Under the chairmanship of Morton Kaplan and Morris Janowitz, the conference provided a forum in which members of the Center for Policy Study and other interested scholars sought to reassess the impact of the military-industrial complex on arms control. In addition, some of the papers derive from the work of the Inter-University Seminar on Armed Forces and Society.

The first section considers the definitions and composition of the military-industrial complex. An analytical essay by Charles Moskos surveys the historical evolution of the concept and relates it to the current debate. In assessing the literature concerning elites and the military-industrial complex, Moskos identifies three elements of the military-industrial complex; a military hierarchy, administrative bureaucracy, and corporate wealth. Each have internal contradictions and represent ambiguous interpretations regarding the military-industrial complex. Furthermore, Moskos notes that the available literature has shifted its focus from societal considerations to military-industrial collusion and the dominance of military officers in the national political system. He concludes that the concept of the military-industrial complex is ultimately incapable of explaining the surgence of opposition to the military-industrial complex which has come to surface on the American scene in recent years.

Both Charles Wolf and Stanley Lieberson analyze the military-industrial complex in terms of linkages between the military and industrial interests. Wolf points to the many complexities involved in identifying the substance of the military-industrial complex at the same time noting that there is no monolithic structure involved. He suggests that contrary to popular thinking, the military-industrial complex is not the all powerful structure when compared to other administrative complexes such as the educational or medical. In fact, according to Wolf, the level of expenditures (defense) "in 1971 represents the smallest percentage of total government expenditures and smallest percent of GNP, that it has represented since the start of the Korean War in 1950." He concludes that the real difficulty is not the military-industrial complex with respect to arms control, but the difficult and complex nature of arms control itself and the realities of the international arena.

Lieberson's empirical analysis of the elitist and pluralistic approaches with respect to the military-industrial complex focuses on determining the degree of dependency of large corporations on military expenditures. Indeed, there "is some evidence that most corporations would be more prosperous if the government shifted to nonmilitary expenditures." Because of the inconclusiveness of this approach in confirming either the elitist or pluralistic theses, Lieberson uses another approach, that of "compensating strategies." This approach is based on the premise that a number of diverse and powerful interest groups will attempt to maximize their gains thereby running counter to the interests of the majority. Using the composition of Senate Committees, Lieberson concludes that within the Congressional committee systems particular geographical leaders are engaged in the "selective pursuit of vested interests" which is particularly visible with respect to expenditures on the military. He concludes that "military expenditures are not currently a vital and necessary prerequisite to general corporate prosperity. Rather, extensive military spending may be viewed as reflecting the operations of an important and powerful interest group in a setting where other legislative issues have an even greater direct bearing on the prosperity of the remainder of the economic segments."

Focusing on one presumed component of the military-industrial complex, the retired military man Albert Biderman, notes that retirees tend to settle in areas in which defense is a major industry and that 50-60 percent of retiree employment is directly dependent upon defense and aerospace industries. Nevertheless, he finds that, with the exception of a few notable personages, most retirees have little impact on decision making within industries which would substantially affect defense contracting. Furthermore, he concludes that "On a national scale, retirees (military) are not now nor are they destined to be numerous enough to be an important economic factor in the employment economy."

Shifting from the analysis of the composition of the military-industrial complex, the second section of the volume explores several key components of the defense structure. Morris Janowitz, in his essay examines the nature of the emerging military establishment and the impact of the all-volunteer military force. He suggests that the end of selective service will have important consequences in both domestic and international areas. The Western European reaction will include a rapid move towards an all-volunteer system in NATO, thereby raising a fundamental question regarding the capability of an all-volunteer system to fulfill international commitments and perform a deterrence role. With respect to U.S. requirements, Janowitz raises two essential questions: "How can the United States forces be redeployed and professionally reorganized so as to articulate with a meaningful and politically responsible foreign policy? How can these all-volunteer forces be recruited, trained, and managed so as to articulate with civilian control and prevent social isolation of the armed forces from the main currents of domestic society?" In discussing these issues, Janowitz re-emphasizes the need for a "military establishment organized and oriented towards a constabulary concept in which it is committed to the minimum use of force and seeks viable international relations, rather than victory, because it has incorporated a protective military posture." Finally, he argues that the character of the all-volunteer force is not solely a military question, but one that stems from the nature of the larger society.

The essays by Seymour Melman and Bruce M. Russett view strategic alternatives as one means of reducing conflict probabilities while controlling the scope of the defense establishment. Melman argues for a greatly reduced defense budget, which will release resources for domestic priorities and still need not reduce the U.S. national security posture. Melman postulates "that the security of a society consists of protection against destruction from without, as well as the well-being of its people. A competent security policy must serve *both* these ends." Although much of his essay is devoted to a detailed accounting of an alternative defense budget, Melman argues that this is presented in terms of what is the most reasonable defense posture rather than simply in terms of cost reduction. He concludes that "the military security of the United States can be completely served with armed forces costing two-thirds less, between $25 and $29 billion."

Russett introduces the concept of a countercombatant strategy which he defines as more than retaliation against nuclear striking forces, but less than a countercity or countervalue strike, which focuses on high accuracy strikes against selected military facilities without much regard to their location, "without inflicting nearly such a high level of civilian casualties as would be implied by a deliberate countercity strategy." The objective is to reduce the enemy's ability to control and use his military forces. By adopting such a strategy, part of the moral dilemma involved in including civilian targets might be resolved, as well as reducing the arms race syndrome. Conflict restraint is created by such a strategy, without requiring reciprocity by the other side, at the same time producing a receptive international environment for arms control.

Arms and conflict control, the subject of the third section of the volume is another central component of the defense effort. In approaching the issue on a general level, Wayne Smith and Jack Merritt argue that answers to arms control measures are highly complex and multi-dimensional involving varying perceptions of political and security interests. In reviewing a number of elements which they assert are essential for successful and realistic arms control negotiations, the essay points out that "arms control rather than being confined to a simple negotiation, is a piece of a process involving force planning and deployment, the totality of international relations, internal economic social pressures, and institutional bargaining." Therefore, according to the authors, to reach agreement on each of these issues, as well as the kind of environment they create requires a common interest and arduous rounds of negotiations and time. Ultimately, it requires a comprehensive examination of the very basis of each political system represented in the negotiating parties followed by a comprehensive agreement. They conclude that "Arms control is fundamentally a political matter.... It requires time and effort to reconcile opposing internal views and for both parties to make necessary political decision to turn a proposal into an equitable and durable agreement."

Appropriately, the fourth and final section of the volume is devoted to problems of decision making in the U.S. executive branch and in its relationships to Congress. In his essay, Adam Yarmolinsky stresses the fact that the President and Congress are primarily responsible for the control of the military and its budgets. In this respect, he argues that the size and shape of the military should be directly related to policy goals. Arms control must be viewed in terms of policy goals and subsequent force structure. Noting that there are two approaches to arms control, procedural and substantive, Yarmolinsky suggests that a substantive approach "is more responsive to budgetary constraints and therefore more easily achievable." He stresses the need for both President and Congress to become more involved in incremental decision-making as a means of achieving control of and giving direction to the defense posture, by making choices in the defense budget. Reviewing the components of this decision-making as timeliness, objectivity, discipline, and information, he concludes that choices in the budget process rests with the political leadership of the country.

Lawrence Korb focuses on the mechanics of the budgetary process within the defense department, and as such provides insights into the arguments advanced by Yarmolinsky. After a detailed review of the budgetary process since the Eisenhower administration, Korb concludes that there has been a process of continual change in the relationships between the Secretary of Defense and the Joint Chiefs of Staff. Much of this process stems from the perceptions of the role of the Secretary of Defense as viewed by the incumbent. On the other hand, the Joint Chiefs of Staff must accept the role allocated to them in the budgetary process by the Secretary of Defense. Korb notes that the Congressional outlook and popular public opinion also affect the relationships between the Joint Chiefs of Staff and the Secretary of Defense. In the main, these relationships have been in accord with the general functioning of the political system. Korb suggests that in the main, the military establishment must accommodate itself to the wishes of the Secretary of Defense and that any disagreement between the two usually is rectified by the President who inevitably sides with the Secretary of Defense. In this respect, the primary responsibility for control and management of the defense establishment rests with the political leadership.

These contributions should serve the purpose of underlining the need for scholarly assessment of the military-industrial complex divorced from ideological rationalizations. Scholarly analyses militate against a number of prevailing popular assumptions, suggest a new research approach, while in some respects they often tend to confirm and reinforce previously held premises. Few would deny the necessity of the defense establishment. However, one must be careful not to equate necessity with dominance. In any case, the complexities of the issue and the pervasiveness of the U.S. defense effort does require a continuing systematic critique not only in terms of specific issues, but of the nature of the political system.

NOTES

This volume was prepared with the secretarial assistance of Paul Hancock, Loyola University of Chicago. Additionally, Phyllis Oman and Tom Ferguson, both of Loyola University provided help in secretarial and research matters.

*Clark R. Mollenhoff, *The Pentagon, Politics, Profit and Plunder* (New York: Pinnacle Books, 1972), p. 57.

Part I

CONCEPTS AND REALITIES

THE MILITARY-INDUSTRIAL COMPLEX: THEORETICAL ANTECEDENTS AND CONCEPTUAL CONTRADICTIONS

Charles C. Moskos, Jr.

The nature of the military-industrial complex has given rise to one of those Great Debates that periodically appear on the American public scene. The issue is not so much the existence of a military-industrial complex -- which few would deny -- but to what degree it is an autonomous entity and to what purposes it is directed. Academic definitions as well as public attitudes range across a variety of positions. At one end are those who see the military-industrial complex as one of many institutional linkages occurring in our society and which was engendered primarily by quite real external threats. Conversely, others see the military-industrial complex as a self-generating structure which is a force for repression at home and abroad. And it is this latter critical view which has come to dominate the rapidly expanding literature on the military-industrial complex and related themes. There is general agreement on the part of such critics that the military establishment is grossly swollen, that military funding has distorted the economy and academia, that the public has been manipulated to foster Cold War attitudes, and that there is a self-serving accomodation between corporate elites, government bureaucrats, and the military hierarchy.

Yet, although often obscured by a common opposition to the ruling structure and contemporary political economy of the United States, there are also profound theoretical differences among the critics as to the causes and prime determinants of the military-industrial complex. Some of the inherent ambiguities in the concept of the military-industrial complex is attested to by the multitude of associated terms used to characterize American society in recent writings in this genre: military-industrial bureaucracy,[1] Pentagon capitalism,[2] state capitalism,[3] military socialism,[4] garrison society,[5] weapons culture,[6] military America,[7] armed society,[8] warfare state,[9] and national security state.[10]

The purpose of this essay is to examine the concept of the military-industrial complex as it has developed and gained currency in the writings of social scientists and public commentators.[11] More concretely, we will (1) inquire into the theoretical antecedents of the military-industrial complex concept; and (2) appraise the current literature by categorizing the prime determinants imputed to the military-industrial complex by various writers. Put most simply, the thesis presented here is that the concept of the military-industrial complex now in vogue, although purportedly a radical critique of the status quo, is in fact a very conventional outgrowth of a pervasive conventional ideology. Rather than being derived from a radical sociological perspective, the concept of the military-industrial complex is consistent with the main trends of American social science and derived from an anti-Marxist sociological tradition in Western thought.

THEORETICAL ANTECEDENTS

It is basic to start any discussion of the military-industrial complex concept with reference to C. Wright Mills' The Power Elite published in 1956.[12] Mills' study is important both because it serves as a benchmark for subsequent writings and because it reflects a distillation of certain major sociological traditions. As is well known, Mills posits a small and unified power elite who control the means of destruction, production, and political power in American society. The power elite consists of a triumvirate of top elites coming from the pinnacles of the corporate, political, and military bureaucracies. They have similar interests in maintaining themselves in power and similar values in matters of public policy. Most emphatically, the power elite is not a ruling class based on ownership of property. Mills specifically states that "ruling class" is a badly loaded phrase implying the political rule of an economic class. For Mills, instead, the term power elite is preferable because it encompasses "political" and "military" determinisms as well as "economic" determinisms.[13]

The power elite model, along with its derivations in the concept of the military-industrial complex, draws heavily on two anti-Marxist schools in classical sociology. The first is the Machiavellian tradition as represented in the writings of Vilfredo Pareto and Gaetano Mosca. The Machiavellian perspective substitutes an elite/mass cleavage for class conflict. The ruling elites (whom the Machiavellians often imbued with power-seeking psychological qualities) are seen primarily as governmental and political leaders -- appointive as well as elected; not as landed or business elites. The other classical approach which arose in opposition to Marxist theory is represented by Max Weber. Centrally concerned with the nature of authority in complex social organizations, Weber saw power arising not only from capital ownership, but also -- and more important -- from occupancy of top positions in governmental bureaucracies. Thus, for both the Machiavellians and Weber, the elite bureaucrat and not the capitalist (or for that matter the general) was the power figure of our time. Although undeniably influenced by Marxist concepts, Mills' The Power Elite remains firmly in the Machiavellian and Weberian mold with its deemphasis of economic class and reliance instead on notions of bureaucratic elites.

A decade and a half before The Power Elite was published, two germinal studies appeared which also combined the Machiavellian and Weberian traditions. On the eve of the outbreak of World War II, Harold Lasswell stated his theory of modern civil-military relations in his concept of the "garrison state."[14] Forecasting a particular form of social organization, the garrison state would be characterized by the militarization of the civil order as the military system became coterminous with the larger society. There would be an ascendancy of the "specialists in violence" and a corresponding obliteration of the distinction between civil and military personnel. In 1941, James Burnham argued in The Managerial Revolution that modern industrialized societies were governed by managers who in effect transcended the means of production because they decided corporate and government policies.[15] Burnham's managerial revolution entailed the rise of top executives (in industry, government, and labor) who would ultimately triumph over propertied classes. For both Lasswell and Burnham, then, there was to be no "ruling class" but rather a ruling group of elite managers, specialists, and bureaucrats.

Contemporary with the publication of The Power Elite was Ralf Dahrendorf's Class and Class Conflict in Industrial Society. Restating the managerial revolution concept, Dahrendorf described the ruling group of the post-industrial state as composed "of the administrative staff of the state, governmental elites at its head, and those interested parties which are represented by the governmental elite."[16] More recent variations on the same theme are found in Suzanne Keller's Beyond the Ruling Class,[17] and John Kenneth Galbraith's The New Industrial State.[18] In all of these studies, despite contrasting political positions, there was a common conceptual ancestry in the ideas of the Machiavellians and Weber. Contrary to Marxist thought, these theorists portrayed post-industrialized societies -- whether capitalist or socialist -- as converging in their basic forms of authority; top leaders emerging out of large-scale public and private enterprises comprised the new ruling groups; and economic class conflict was being superceded by the bureaucratization of elite structures.[19]

The point here is simply that the basic theoretical premises underlying C. Wright Mills' The Power Elite were not in themselves departures from prevailing sociological thought. Quite the contrary, the central conceptual thrust of The Power Elite was essentially in accord with the basic formulations of earlier and later non- if not anti-Marxist social scientists.[20] In particular, the invocation of notions such as mass society and elite groups served to undercut explanatory concepts based on economic relations of classes to the means of production. Moreover, this aversion to use a class analysis has similarly characterized the subsequent development of the concept of the military-industrial complex.

MILITARY-INDUSTRIAL DETERMINANTS

It is not surprising that interpretations of the causal factors in the operation and makeup of the military-industrial complex are many. At first glance it would appear that the literature on the military-industrial complex is too variable to categorize readily. From among the diverse accounts, however, we can distinguish -- with some overlap to be sure -- three recurring and analytically exclusive themes as to the root equality of the military-industrial complex. For purposes of exposition, we can refer to these three determinants as deriving from (1) the military hierarchy; or (2) the administrative bureaucracy; or (3) corporate wealth. From a conceptual viewpoint, the prevailing interpretation of the military-industrial complex is most closely associated with those stressing military factors, somewhat less so for those focusing on bureaucratic variables, and least of all for researchers who deal with economic determinants.

The Military Hierarchy

The truly novel feature of The Power Elite was Mills' introduction of the military "warlords" as constituent members of the ruling elite of the United States. Never before had uniformed officers been so characterized in American society.[21] This portrayal of the military found an echo in J.R. Swomley's The Military Establishment which squarely placed the blame for the militarization of American society on the "permanent officer group."[22] But it was not until the contemporary period when the war in Indochina engendered a virulent anti-militarism within America's intellectual and academic circles that the nefarious role of the uniformed military became a leitmotiv in much of the military-industrial complex literature.

In The Economy of Death, Richard J. Barnet thus states: "By the end of the Johnson Administration, the uniformed military had acquired considerable independent political power."[23] Sidney Lens whose The Military-Industrial Complex includes a panoply of vested interests, nevertheless, points to the Pentagon as "the fountainhead" of American imperialism.[24] Robert Heilbroner concludes "...that the military establishment constitutes itself as a self-contained entity, capable of impressing its views and imposing its will not only on the civil establishment to which it pays ritual obeisance, but over a section of the economy in which the language of private enterprise is merely a fiction to hide its absolute authority."[25] John Kenneth Galbraith holds: "In my view the Services, not their industrial suppliers, are the prime wielders of this power."[26] Indeed, Galbraith explicitly reverses Marxist causality and refers to military suppliers as "captive contractors."[27]

Specifying the military as the main culprits in the military-industrial complex is not confined by any means to academics. Thus former Marine Commandant David Shoup argues that inter-service rivalry plays a determining role in weapons acquisitions and American interventions abroad.[28] Senator William J. Fulbright's The Pentagon Propaganda Machine forcefully castigates senior officers and Defense Department officials for brainwashing the American public into acceptance of military values and priorities.[29] Similarly focusing on the military as the major agents of the military-industrial complex are those accounts dealing with the role of retired military in defense industries and in reports of military waste and collusion between high-ranking officers and armaments manufacturers.[30]

Running through the arguments of the military determinists is the fear that the senior command structure has been slipping out from under civilian control. The emphasis from this perspective on the military-industrial complex is definitely on the military.[31] Advocates of this viewpoint are usually self-conscious political liberals who eschew radical attacks on the American social system. They reflect the sentiment that once the armed forces are returned to their proper (and much reduced) role, the system will right itself.

The Administrative Bureaucracy

Deriving almost directly from Machiavellian elitism and Weberian bureaucracy via the "managerial revolution" and the "power elite" is that military-industrial complex literature which sees the causal factor vested in an autonomous bureaucracy of non-elected administrators. This viewpoint is perhaps most persuasively presented in Pentagon Capitalism by Seymour Melman.

Briefly, Melman's thesis is that the military-industrial complex of the Eisenhower era has been supplanted by a "state-management" decision system introduced by Secretary of Defense McNamara in the early 1960s. Melman argues that the military-industrial firm is no longer a private firm in that corporate decisions are now in the hands of government managers. "The state management has also become the most powerful decision-making unit in the United States government. Thereby, the federal government does not 'serve' business or 'regulate' business. For the new management is the largest of them all. Government *is* business. That is state capitalism."[32] (Italics in the original.) Significantly, defense strategy has become a means to an end, the end being greater domestic power and influence. For Melman, then, the primary goals of the Defense Department are the expansion of its own power within American society.

A common corollary of administrative bureaucratic determinism is to portray governmental elites as propelled by an "institutionalized power lust,"[33] or "executive egoism,"[34] or "bureaucratic machismo."[35] The absence of class analysis in this train of thought is manifest when Irving Louis Horowitz reviews The Pentagon Papers not in terms of a military-industrial complex dominated by economic oligarchs, nor even of a power elite, but by specifying a ruling group in American society of a "non-elected political elite."[36] Carrying this mode of analysis one step farther, Robert W. Tucker describes the compulsion for imperialist America as the quest for power itself and not because of any internal capitalist dynamic.[37] Somewhat more involved is the thesis presented by Pilisuk and Hayden that the value premises of American elites are in fact not that much different from the mass of the population.[38] "American society is a military-industrial complex" because of the pervasiveness of a set of cold-war-counter-revolutionary "core beliefs." For Pilisuk and Hayden, that is, precisely because of the unquestioned acceptance of these core beliefs, it is not necessary to invoke a class analysis to understand the dominance of the military-industrial complex in American society.

Corporate Wealth

Contrasting with both those who emphasize military dominance of administrative bureaucrats is the Marxian perspective which interprets American society in terms of class structure and corporate wealth. In Marxian analysis, military determinism is only a misguided scapegoating of a superstructural entity, while administrative determinism erroneously reverses the independent-dependent relationship between government official and corporate wealth. Indeed, the whole concept of a military-industrial complex is incompatible with a Marxian framework as was also the earlier case with Mills' thesis of a power elite.

Paul A. Baran and Paul M. Sweezy in their Monopoly Capital make the central point that the huge American military machine serves the capitalist purpose of maintaining prosperity at home while fighting socialism abroad.[39] Even more directly David Horowitz states: "The locus of power and interest (of the "ruling class") is the giant corporations and financial institutions which dominate the American economy, and moreover, the economy of the entire Western world."[40] William Appleman Williams in his Roots of the Modern American Empire sees the Cold War greatly antedating the military-industrial complex and traces its origins to the nature of the integral expansion of capitalist society in America.[41] Victor Perlo's Militarism and Industry,[42] and Gabriel Kolko's The Roots of American Foreign Policy[43] and The Politics of War[44] give detailed documentation of how the corporate elite control those aspects of the governmental and military machinery which are essential to capitalist interests. Common to such Marxian analyses is the characterization of the military and the governmental bureaucracy playing a minor role in American corporate and imperialist decision making.[45]

Emphasis on the determining role of the corporate rich in shaping basic cold-war-defense expenditures is not exclusively a Marxist perogative, however.[46] Adopting an essentially non-Marxist but radical framework are several studies which similarly conclude that an aristocratic economic elite sets the fundamental policies of America's defense establishment. This is the conclusion of G. William Domhoff's survey of America's power structure.[47] Stephen A. Cobb views the contemporary corporate rich of the defense economy being incorporated into the ruling class in a manner akin to the integration of the "robber barons" into the national elite after the Civil War.[48] Ferdinand Lundberg in a scathing critique of C. Wright Mills holds that the military hierarchy, governmental leaders, and corporate managers are little more than handmaidens of the "super-rich."[49] For Lundberg the significant units of power in the United States are wealthy families.

Even in this brief overview of the concept of the military-industrial complex it can be seen that the different theoretical emphases as to the principal determinants are correlated -- albeit loosely and with some overlap -- with rather distinct positions on the political spectrum. The dominant thrust in the literature which stresses the autonomous role of the military hierarchy finds its proponents among liberal spokesmen; emphasis on the administrative bureaucrats seems to be chiefly a characteristic of those adhering to a kind of New Left persuasion; while attributing prime determinancy to corporate wealth finds its adherents among Marxist writers.

To recapitulate and conclude, then, the concept of the military-industrial complex is not without its internal contradictions. Customarily regarded as the antipode of conventional bourgeois sociology, the concept of the military-industrial complex is quite in the mainstream of the development of American social science. Purportedly a radical critique of the status quo, the overriding thrust of the military-industrial complex literature has been to shift attention away from societal and structural considerations to revelations of military-industrial collusion and corruption and the alleged dominance of military officers in the national power system. Presented as a mode of analysis which is both current and relevant, the concept of the military-industrial complex is ultimately incapable of explaining the surgence of opposition to the military-industrial complex which has come to surface on the American public scene in recent years.

NOTES

[1] John Kenneth Galbraith, How to Control the Military (New York: Signet, 1969).

[2] Seymour Melman, Pentagon Capitalism (New York: McGraw-Hill, 1970).

[3] Ibid.

[4] Richard J. Barnet, The Economy of Death (New York: Atheneum, 1969).

[5] Vernon K. Dibble, "The Garrison Society," in Seymour Melman, ed., The War Economy of the United States (New York: St. Martin's Press, 1971), pp. 179-186.

[6] Richard E. Lapp, The Weapons Culture (New York: Norton, 1968).

[7] Robert Heilbroner, "Military America," New York Review of Books, Vol. 15, No. 2 (July 23, 1970), pp. 5-8.

[8] Tristram Coffin, The Armed Society (Baltimore, Md.: Penguin, 1964).

[9] Fred J. Cook, The Warfare State (New York: Macmillan, 1962).

[10] Robert Borosage, "The Making of the National Security State," in Leonard S. Rodberg and Derek Shearer, eds., The Pentagon Watchers (Garden City, N.Y.: Doubleday, 1970), pp. 3-63.

[11] For other efforts to appraise the conceptual development of the military-industrial complex, see Jerome Slater and Terry Nardin, "The Concept of a Military-Industrial Complex," paper presented at the Sixth Annual Conference of the Inter-University Seminar on Armed Forces and Society, Chicago, Illinois, Nov. 18-20, 1971; and Max L. Stackhouse, "Competing Definitions of the Military-Industrial Complex," in his The Ethics of Necropolis (Boston: Beacon Press, 1971), pp. 25-42.

[12] C. Wright Mills, The Power Elite (New York: Oxford University Press, 1956).

[13] Ibid., p. 277.

[14] Harold D. Lasswell, "The Garrison State," American Journal of Sociology, Vol. 46 (January, 1941), pp. 455-468. In this early version of his garrison-state concept, Lasswell identified the specialists in violence with military officers. However, in a later formulation, Lasswell included internal security and police forces as well as the military in the ruling elite of the garrison state. See Lasswell, "The Garrison-State Hypothesis Today," in Samuel P. Huntington, ed., Changing Patterns of Military Politics (New York: Free Press, 1962), pp. 51-70.

[15] James Burnham, The Managerial Revolution (New York: John Day, 1941). The notion that groups with managerial and/or technical skills would ultimately wield more power than functionally "superfluous" capitalists in advanced industrialized societies has precedents in earlier American economic thought. See Thorstein Veblen, The Engineers and the Price System (Clifton, N.J.: Kelley, 1921); and A. A. Berle, Jr. and Gardiner C. Means, The Modern Corporation and Private Property (New York: Macmillan, 1933).

Ironically enough, despite the deemphasis of economic class in The Power Elite, C. Wright Mills (with Hans H. Gerth) strongly condemned Burnham's The Managerial Revolution for ignoring the overriding importance of class realities and property ownership. See Mills and Gerth, "A Marx for Managers, Ethics, Vol. 52, No. 2 (January, 1942) reprinted in Irving L. Horowitz, ed., C. Wright Mills (New York: Ballantine, 1965), pp. 53-76.

[16] Ralf Dahrendorf, Class and Class Conflict in Industrial Society (Stanford: Stanford University Press, 1956), p. 303.

[17] Suzzane Keller, Beyond the Ruling Class (New York: Random House, 1963).

[18] John Kenneth Galbraith, The New Industrial State (New York: Signet, 1967).

[19] For a trenchant critique of these premises of "bourgeois" sociology, see H. Frankel, Capitalist Society and Modern Sociology (Surrey, England: Laurence and Wishart, 1970).

[20] Further amplification of this view is found in the Marxist critiques of The Power Elite. See, for example, Paul M. Sweezy, "Power Elite or Ruling Class," and Herbert Aptheker, "Power in America," in G. William Domhoff and Hoyt B. Ballard, compilers, C. Wright Mills and the Power Elite (Boston: Beacon, 1968).

[21] Mills' positing a military determinancy in The Power Elite had, of course, a conceptual antecedent in the ascendancy of the "specialist in violence" in Lasswell's garrison-state theory. Moreover, he was influenced by Bruce Catton, The War Lords of Washington (New York: Harcourt, Brace, 1948); an account of the role of the military in national decision making during World War II. On this point see the discussion in G. William Domhoff, Who Rules America? (Englewood Cliffs, N.J.: Prentice-Hall, 1967), pp. 118-120.

Significantly, two other books also published the same year as The Power Elite similarly alerted their readers to the dangers of militarism in the United States: Walter Millis, Arms and Men (New York: Putnam's Sons, 1956); and Arthur A. Ekrich, The Civilian and the Military (New York: Oxford University Press, 1956). But it was The Power Elite which brought into currency the notion of the military as a prime power group in American socio-ideological thought. Mills' fascination with the autonomous role of the military is also evident in Han H. Gerth and C. Wright Mills, Character and Social Structure (New York: Harcourt, Brace, and World, 1953), pp. 223-229, 284-287.

Cf. Morris Janowitz, The Professional Soldier (Glencoe, Ill.: Free Press, 1960), p. viii: "The military profession is not a monolithic power group . . . Instead, the military profession and the military establishment conform to the pattern of an administrative pressure group, but one with a strong internal conflict of interest. It is a very special group because of its immense resources, and because of its grave problems of national security."

[22] J. M. Swomley, Jr., The Military Establishment (Boston: Beacon, 1964). A similar pre-Vietnam War account of the American military's usurpation of civilian rights is found in Cook, op. cit.

[23] Barnet, op. cit., p. 81.

[24] Sidney Lens, The Military-Industrial Complex (Philadelphia: Pilgrim, 1970), p. 39.

[25] Heilbroner, op. cit., p. 6.

[26] Galbraith, How to Control the Military, op. cit., p. 10.

[27] Ibid., p. 33.

[28] David Shoup, "The New American Militarism," The Atlantic Monthly, April, 1969. See also James A. Donovan, Militarism U.S.A. (New York: Scribner's Sons, 1970).

[29] J. W. Fulbright, The Pentagon Propaganda Machine (New York: Liveright, 1970). See also Rodberg and Shearer, eds., The Pentagon Watchers, op. cit.

[30] See, for example, William Proxmire, Report from Wasteland: America's Military-Industrial Complex (New York: Praeger, 1970); and C. Merton Tyrell, Pentagon Partners: The New Mobility (New York: Grossman, 1970). Cf. John Stanley Baumgartner, The Lonely Warriors (Los Angeles: Nash, 1970); unique because it is a defense of the weapons procurement system. For a balanced treatment of the available data dealing with retired officers in defense industries, see Albert D. Biderman, "Retired Military Within and Without the Military-Industrial Complex" in this volume.

It should be noted that the contemporary emphasis which deals with military-industrial collusion in armaments procurement while having superficial resemblance to the "merchants of death" literature of the 1930's has an entirely different conceptual premise. The 1930's literature rather than positing a military-industrial complex focused instead almost entirely on rapacious arms manufacturers. See, for example, George Seldes, Iron, Blood, and Profits (New York: Harper, 1934); Helmuth C. Engelbrecht and Frank C. Hanighen, Merchants of Death (New York: Dodd, Mead, 1934); and John E. Wiltz, In Search of Peace: The Senate Munitions Inquiry, 1934-1936, (Baton Rouge: Louisiana State University Press, 1963).

[31] In addition to previously mentioned sources, see also Erwin Knoll and Judith Nies McFadden, eds., American Militarism 1970 (New York: Viking Press, 1969). A milder version (although not consistently) of the independent role of the armed forces in the operation of the military-industrial complex is Adam Yarmolinsky, The Military Establishment (New York: Harper and Row, 1970).

A provocative theoretical formulation which relates variations in the military's role in civil affairs to stages in national development is Ritchie P. Lowry, "To Arms: Changing Military Roles and the Military-Industrial Complex," Social Problems, Vol. 18, No. 1 (Summer, 1970), pp. 3-16.

[32]Melman, Pentagon Capitalism, op. cit., p. 2. An excellent exposition of the Melman thesis is a relation to current studies of the defense establishment in Stephen J. Cimbala, "New Myths and Old Realities: Defense and Its Critics," World Politics, Vol. 24, No. 1 (Oct. 1971), pp. 127-157.

[33]Melman, op. cit., p. 4.

[34]Irving Louis Horowitz, "The Pentagon Papers," Trans-Action, September, 1971, p. 38.

[35]Richard J. Barnet, "The Game of Nations," Harper's Magazine, November, 1971, p. 55.

[36]Horowitz, op. cit., p. 38. Horowitz has consistently assigned a primary role to the "civilian militarist" who formulates national policy rather than to corporate executives or senior military officers. See Irving L. Horowitz, The War Game (New York: Ballantine, 1963). See also the emphasis given by Horowitz to government bureaucrats in his Three Worlds of Development, second edition (New York: Oxford University Press, 1972).

[37]Robert W. Tucker, The Radical Left and American Foreign Policy (Baltimore: Johns Hopkins University Press, 1971).

[38]Marc Pilisuk and Tom Hayden, "Is There a Military-Industrial Complex that Prevents Peace?" in Robert Perrucci and Marc Pilisuk, eds., The Triple Revolution (Boston: Little, Brown, 1971), pp. 73-94. A similar appraisal of the military-industrial complex in terms of the primacy of core values in the political culture of the United States is found in Stackhouse, op. cit.

[39]Paul A. Baran and Paul M. Sweezy, Monopoly Capital (New York: Monthly Review Press, 1966). Cf. Albert Szymanski, "Military Spending and Economic Stagnation," paper to be presented at the 1972 meetings of the American Sociological Association, New Orleans, La.; Szymanski argues that the Baran and Sweezy model is inaccurate in that monopoly capitalism has a flexibility not necessarily geared to military expenditures.

[40]David Horowitz, "Introduction," in his edited Corporations and the Cold War (New York: Modern Reader, 1970), p. 11.

[41]William Appleman Williams, Roots of the Modern American Empire (New York: Random House, 1970). Indeed, Williams writes in another context that "...the cold war actually began with the triumph of laissez-faire capitalism over the more organic political economy of mercantilism, and can be dated by the publication of Adam Smith's Wealth of Nations in 1776." (!)

Williams, <u>New York Review of Books</u>, Oct. 2, 1971, p. 3.

[42] Victor Perlo, <u>Militarism and Industry</u> (New York: International Publishers, 1963).

[43] Gabriel Kolko, <u>The Roots of American Foreign Policy</u> (Boston: Beacon, 1969).

[44] Gabriel Kolko, <u>The Politics of War</u> (New York: Random House, 1970).

[45] However, Harry Magdoff, "Militarism and Imperialism," <u>Monthly Review</u>, Vol. 21, No. 9 (February, 1970), pp. 1-14, holds that militarism is interwoven with imperialism and that both phenomena must be accounted for in Marxian terms.

It should also be noted that a Marxian perspective does not necessarily assume a unified ruling class. Perlo, <u>op. cit.</u>, shows how different elements of the corporate structure have varying levels of intimacy with the military-industrial complex. Indeed, Soviet analysts interpreted the publication of <u>The Pentagon Papers</u> as an outcome of a struggle between opposing elements within the military-industrial complex; one faction concerned with conventional weapons and their use in Indo-china, the other supplying strategic weapons and seeking disengagement from South-east Asia. From <u>Literaturnaya Gazeta</u> cited in <u>The New York Times</u>, July 15, 1971, p. 6.

At a more crass level, the import of narrow economic self-interest on the part of top leaders must not be so easily dismissed as is the custom in the military-industrial complex literature. For example, evidence from the Dominican Republic intervention in 1965-66 shows that key administrative decision-makers were intimately connected with large holdings in the Caribbean. See Fred Goff and Michael Locker, "The Violence of Domination: U.S. Power and the Dominican Republic," in Irving L. Horowitz, et al., eds., <u>Latin American Radicalism</u> (New York: Vintage, 1960), pp. 249-291.

[46] A quite different and potentially fruitful approach to the study of the military-industrial complex--though still within the broad framework of economic determinants--is found in Stanley Lieberson, "An Empirical Study of Military-Industrial Linkages," <u>American Journal of Sociology</u>, Vol. 76, No. 4, January, 1971, pp. 562-584 (reprinted in this volume). Briefly Lieberson's theory of "compensating strategies" holds that extensive defense expenditures reflect not so much a coordination of interests among major corporations, but rather that the big gainers of the defense industry have caused only marginal diminishment of the gains of non-defense industries. Because each industry concentrates its strength in its own area, Lieberson concludes that

there is no necessity to posit a conspiracy to produce high levels of defense spending.

[47]Domhoff, Who Rules America?, op. cit.

[48]Stephen A. Cobb, "The Military-Industrial Complex: Elite Domination or the Pork Barrel?", paper presented at annual meetings of the International Studies Association, Dallas, Texas, March 15-18, 1972.

[49]Ferdinand Lundberg, The Rich and the Super-Rich (New York: Bantam, 1968). See especially pp. 543-553 and 950-954 where Lundberg frontally attacks the concept of a power elite which he correctly interprets--C. Wright Mills' disclaimers to the contrary--as a restatement of the managerial revolution thesis.

MILITARY-INDUSTRIAL SIMPLICITIES,
COMPLEXITIES AND REALITIES

Charles Wolf, Jr.

One of the central questions to which this symposium is addressed is whether and to what extent the military-industrial complex is an obstacle to arms control? As posed, the question assumes that there <u>is</u> something that can accurately be referred to as a "military-industrial complex," and that the term connotes something clearly (and similarly) understood by readers or listeners. The first assumption is at least debatable; the second assumption much less debatable because much more assuredly wrong.

Consider three different meanings frequently ascribed to the term, tacitly or explicitly.

The first I will characterize as a <u>primitive monolithic</u> (PM) view of the MIC.[1] According to its author, the MIC is "an outgrowth of a new ... concept of national purpose -- global expansion," whose effect has been "to <u>weld together</u> those elitist elements at home which have a stake in militarism -- the Armed Forces, a group of legislators, industrialists, government officials, the labor hierarchy, an important segment of academia -- into ... the military-industrial complex." In its behavior, the MIC seeks to "manufacture a public stance of hard-line anti-communism." It also "withholds information and misinforms the public ... (and) inhibits the process of dissent...". The MIC, "far from assuring peace, ... develops a momentum for war. The very preparation for war has become an independent factor in promoting it."

A milder version of the same view is advanced, though not uniformly, in Adam Yarmolinsky's recent study, which speaks of "the coincidence of self-interest, in regard to military expansion, between Congress and the military-industrial establishment."[2]

A second meaning, which I will characterize as a <u>primitive pluralistic</u> (PP) view of the MIC, was recently advanced by Soviet writers in the <u>Literaturnaya Gazeta</u> to explain publication of the Pentagon papers by certain American newspapers. According to this view, the MIC's industrial arm has two components: one principally concerned with producing conventional arms used in Indochina; the other consisting of the chief suppliers of strategic weapons. The two components have differing interests, notably reflected in the competition between them for defense expenditures. This conflict within the MIC, combined with a further split between the conventional-weapons part of the MIC and the "civilian-industrial complex," led to publication of the Pentagon papers, according to the PP view.[3] The aspect of PP that is germane to the subject of this seminar is the perception that the MIC is divided, that it has splits within it, and that there are some (in this case, the "civilian" complex) effective counterweights against it.

I don't mean to caricature these views. They are held, and quite firmly, by a number of people. Yet they seem to me quite wide of the mark, simplistic and misleading formulations of phenomena which are more complex and diffuse. Let me suggest some of the complexities, embodying them in a third view that seems to me more nearly correct.

The so-called MIC consists of many different "turfs," with differing and often conflicting interests, indicators, perceptions, managers, and behavioral patterns. Three turfs are occupied by the <u>military services.</u> Their primary focus is on national security, among the country's national goals. The services are also concerned with preserving and, if possible, raising their budgets, and their budget <u>shares</u> relative to the other services. There is thus competition, sometimes excessive, between them, not only for budgets, but for roles and missions which get reflected in budget shares: competition between the Air Force and the Navy for strategic missions as well as for tactical air missions; between the Army and the Air Force for close air support; between the Army and the Marines for air mobile <u>versus</u> amphibious forces, respectively. This rivalry can be perverse in various ways, not excluding log-rolling. It can also be benign, a source of discipline tending to keep each of the services from overstating a position, or making an excessive claim on resources for research and development, or weapon systems acquisition. Both the perversities and benefactions are parts of the complex reality of the MIC. Concentrating on either one to the exclusion of the other is simplistic and misleading.

Another part of the so-called MIC, <u>defense industry,</u> is itself differentiated along lines not excluding, but not confined to, those suggested by PP. Defense industry includes aerospace industrial firms, for which a major share of total sales is to the Defense Department; large conglomerates, often also in high-technology fields, but usually with only a minority of total sales to the DOD; and various other firms, including small suppliers of components, as well as manufacturers of small arms, and segments of the automotive industry for whom defense procurement is a negligible part of total business. The occupants and management of the industrial "turf" are assuredly concerned with contracts, sales, and profits (although, as the recent example of Lockheed suggests and from other data we will discuss later, their concerns often yield quite mixed and discouraging results). Their interests also include pushing technology forward, particularly where there may be spillover benefits to be realized in the civilian sector, as in the case of computers and jumbo-jet aircraft technology.

The interests encompassed within the defense industrial turf are often in sharp competition with one another. Competitive bidding on development contracts is one evident example, less evident, however, on procurement contracts. Lockheed's recently precarious financial predicament provided other examples. Several competing firms in the aerospace industry publicly urged that the government not provide financial assistance to bail Lockheed out. At least one offered to provide a comparable aircraft for those airlines that had placed orders for the L-1011, if Lockheed went under -- thereby hoping to remove this part of the Lockheed case for emergency governmental assistance, and thereby remove Lockheed as a competitor!

Another of the MIC's turfs is occupied by those Senators

and Congressmen who exercise a major influence in the Congressional authorization and appropriations committees responsible for military affairs. That many of them are keenly mindful of the interests of their local constituencies, as well as of the power that they themselves exercise, does not gainsay the genuine concerns they feel for the security and well-being of the country, nor the considerable knowledge and expertise that they have acquired concerning national security matters. Their concerns and motives are, in other words, mixed, though not necessarily more so than are those of the rest of us. Sometimes the interests and concerns of these legislative leaders accord with those of the military services and of defense industries. And sometimes divergences arise. Examples of divergences between the Congressional leadership and some of the military services include the F-111, the Fast Deployment Logistics Ships, the recently accelerated rate of military pay increases, and the allocation of resources for maintaining local military bases rather than for procurement or other uses.

One can adumbrate other turfs that are usually closely associated with the MIC, notably various private research institutions, and individual members of the scientific and academic community. (As a slight digression, it may be worth mentioning that among the individual and scientific experts testifying in connection with the SAFEGUARD ABM system, there were as many consultants of the Department of Defense who were opponents as were supporters of the Administration's 1969 proposal!)

In any event, even this brief sketch is sufficient to support a few central points about the complexity of the MIC:

1. Its structure is highly pluralistic. The MIC abounds with separate turfs, that have frequent conflicts with one another, and provide rivalry and competition as well as convergence or collusion.

2. Occupants and managers of these turfs are concerned with national welfare and national security interests as well as with local and constituent interests; nor do they perceive any serious conflict between the two.

The reality of the MIC is complex not simple, pluralistic not monolithic, sometimes effective and potent, sometimes ineffective and impotent, no less motivated by concern for national interests than its critics, nor less motivated by a mixture of other motives than its critics.

In support of these complexities about the MIC, let me cite a few numerical realities by way of further refutation of the simplistic views, PM and PP, about the MIC.

Consider three major public issues of the last few years

that were conceived as involving the deepest interests and commitments of the MIC:

1. The SAFEGUARD ABM deployment in 1969;

2. The Cooper-Church Amendment, barring funds for military operations in Cambodia in June 1970; and

3. Appropriations for development of the supersonic transport in December of 1970.

All three were regarded as symbolic of the interests and influence of the MIC. With respect to two of the three issues, the position putatively favored by the MIC won in the Congressional voting. An interesting question arises as to what would one have predicted about the voting by State Congressional delegations, based on adopting as a prior hypothesis either the Primitive-Monolithic or Primitive-Pluralistic views of the MIC?

It seems reasonable that one would have predicted, on the basis of these views, that the proportion of Senate and House votes in support of the putative MIC position on each of these issues would have been closely associated with the relative size of Defense Department contracts by State. This prediction would assuredly follow from the PM view for all three issues. It would also seem to follow from the PP view, at least in the case of the SST.

What do the data show?

If one ranks the States in order of dollars of defense contract awards per capita (for 1969), and compares that ranking with separate rankings of the States in accordance with the percentage of their combined House and Senate votes: (a) <u>for</u> the SAFEGUARD deployment; (b) <u>against</u> Cooper-Church; and (c) <u>for</u> the SST appropriation, one finds that the rank order correlation coefficients are -.12, .1 and .1, respectively. None of these correlation coefficients is significantly different from 0!^{4} The results provide no evidence of a relationship between defense expenditures by States and Congressional voting, on these key issues.

A second set of quantitative data, also inconsistent with the simplistic views of the MIC discussed earlier, and with the exaggerated rhetoric that is sometimes bandied about concerning the arms "race," relates to U.S. defense budget expenditures.

In fiscal year 1961, the last year of tight Eisenhower defense budgets, U.S. defense outlays were about $45 billion, which was approximately 8.8 per cent of the GNP, and a little over 44 per cent of total Federal government expenditures. In 1969, defense expenditures had risen to nearly $79 billion, about 8.7 per cent of GNP, and nearly 42 per cent of total Federal outlays. For fiscal year 1971, defense expenditures were down to $72 billion, about 7 per cent of GNP, and less than 36 per cent of total Federal budget expenditures. Moreover, in constant 1961 prices, defense expenditures in fiscal 1971 were about $55 billion, or just $10 billion more than they were in 1961. And the budget for <u>strategic</u> offense and defense forces in 1971 was only <u>half</u> the 1961 amount, in constant dollars.

For the current fiscal year 1972, defense outlays are expected to rise to about $76 billion. The increase is equivalent to the extent of inflation in the economy as a whole. Hence, in constant prices the real value of FY'72 outlays is about the same as 1971, and the proportion of GNP is slightly <u>below</u> the preceding year.

Thus, while the level of defense expenditures in 1971 is still extremely large, it represents the smallest percentage of total government expenditures and the smallest percentage of GNP, that it has represented since the start of the Korean War in 1950. Even more important, the rates of change are negative and substantial. The military share of government and total resources has gone dramatically <u>downward</u> at the same time as the amount and the shares devoted to health and education have gone dramatically upward, from less than $2.5 billion or 2.5 per cent of federal government expenditures in 1961, to $23 billion or 11.5 per cent of total government expenditures in 1971.

To say, as some do, that military expenditures are "out of control," or that the MIC is a powerful and effective obstacle to limiting arms expenditures, is inconsistent with these data.

A third set of data, that refutes some of the popular imagery about the MIC, is contained in a recent report to the Congress by the General Accounting Office on defense industry profits.[5] The GAO Study showed that profit rates, before Federal income taxes, realized on equity capital investment (which excludes facilities contributed by the government) was almost exactly the same for 32 randomly selected large defense contractors doing more than 10 per cent of their total business with the Defense Department, as for 13 randomly selected contractors doing less than 10 per cent of their total business with the Defense Department. The weighted averages for the 1966-1969 period were 22.7 per cent and 23.1 per cent, respectively.[6] The result nicely accords with what economic theory would suggest, and is sharply at variance with exaggerated rhetoric about high profits realized in defense contracting.

I have so far considered some simplicities, several complexities, and a few numerical realities pertaining to the MIC. Let me now turn to the question of obstacles in the way of arms control, and the MIC's role among these obstacles. The position I am going to advance has three components:

a. First, some degree of arms limitation is already in operation -- the data on defense expenditures presented in the preceding section strongly suggest that "arms race" metaphors are far from the reality that has obtained in the last half dozen or ten years;

b. There are numerous and substantial obstacles in the way of more stringent arms limitations -- not infinite and not insuperable, just numerous and large;

c. Among these obstacles, the role of the MIC is distinctly limited. Emphasizing it with colorful rhetoric and slogans does a disservice to arms limitations, because it turns attention away from the major obstacles on which serious work and effort are needed, including the Soviet Union's own MIC, and whether and how it can be influenced. In fact, one of the extraordinary aspects of the MIC issue is the extent to which sloganeering and invective about it provides an escape from, and a substitute for, careful work on arms limitations problems.

What are some of these major issues and problems?

At the core of the arms limitation problem is the relationship between U.S. behavior with respect to weapons development, deployment, operations, and budgets, and that of the Soviet Union, and of the Chinese. Consider three fundamentally different models.[7] One model postulates positive interaction:

the more the U.S. does with respect to one or more of these variables, the more the Soviets are likely to do, or/and the Chinese; and the less, the less. (Further, the interactions may be contemporaneous or lagged.)

The second model is autonomous: what the Soviets or/and the Chinese do in the military field depends largely on technological opportunities, organizational and bureaucratic continuity, competing resource claims at home, and the scale and composition of economic growth. Consequently, changes in U.S. defense efforts will have little or no effect.

A third model implies negative interaction: if the United States does too little, with respect to offensive or defensive capabilities, or strategic or general purpose forces, the Soviets may be stimulated to expand their own efforts so as to move from "parity" or "sufficiency," to "superiority," perhaps for reasons of political as much as military advantage.

Other formulations are also possible, involving elements of all three of these models, with some degree of simultaneous interaction between U.S. behavior and those of either or both the Soviets and the Chinese. The issue is still further complicated by the fact that differing relationships may apply to different military activities. For example, a different relationship may apply to R&D activities and outlays, from that which applies to the deployment of strategic offense or defense systems, to general purpose forces, or to defense budgets as a whole. Choosing among the different relationships requires a great deal of serious study and analysis, although in fact very little analytical effort has been devoted to this problem. It seems to be the case that some evidence can be cited in support of several of these quite different models, especially with respect to different parts of our own military efforts and those of the Soviet Union.

The important point to make is that, without a better understanding of these relationships, as well as more accurate and timely information-gathering and analysis to keep constant track of whether and how these relationships may change, we are ill-prepared to pursue arms limitations very far. In the absence of understanding, good intentions may have no effects, or even perverse effects. Critics of the American MIC seldom ponder whether and how the Soviet MIC might be influenced, especially if it has the characteristics they attribute to our own.

A second major problem concerns the pairing or sequencing of limitations on strategic offensive and defensive systems. Should an effort be made to obtain a limitation agreement with respect to ABM, with or without limitations on strategic offensive weapons, or in what proportions? This is one of those paradoxical situations where stricter <u>limitations</u> may or may not be equivalent to more effective arms <u>control.</u> For example, in a crisis situation, it can be argued that there is greater stability (and hence "control") if both sides' offensive forces are reasonably well protected, either actively or passively, so that neither side is tempted to strike first. A situation in which both sides have some defensive, as well as offensive, forces may thus be more stable and "controlled" than one in which both sides have only offensive forces. And a situation in which one side has <u>both</u> offensive and defensive forces, and the other side has <u>only</u> offensive forces, may be risky for both, as well as acutely uncomfortable for that "other" side.

Furthermore, the incentives toward nuclear proliferation -- which is a major potential source of arms <u>decontrol</u> -- by other countries (for example, India and/or Pakistan or Japan) might be increased if the U.S. and the Soviets were to confine their strategic capabilities to offensive forces only. Again, the risks entailed may be worth taking, particularly if one is reasonably convinced of the positive interaction model described earlier. But at least one would like to be fully aware of what the risks and the benefits of moving in this direction are. And a clearer awareness requires a lot of hard work and careful study.

There are many other difficult substantive problems

affecting the type and extent of arms control that should be sought in the strategic area. For example, should the United States, by joint agreement or even unilaterally, place less reliance on a three-pronged strategic force (the so-called "triad" consisting of land based missiles, strategic bombers, and missile carrying submarines), and instead limit the U.S. strategic force to only one of these components?

Enthusiastic advocates of arms control sometimes answer this question by advocating a one-pronged force on the grounds that this would be more "limited," and hence presumably better. But would it? Each component of the triad has different system characteristics. Each differs in the accuracy with which it can hit targets, and in its vulnerability to attack. Moreover, each component of the triad presents different problems and uncertainties for an attacking force, thereby complicating the attacker's task, and reducing the incentive to attack. Further, a "monad" instead of the triad would be more vulnerable to the risk of a technological breakthrough that might critically diminish the effective deterrence that such a stripped force could provide. Indeed, the incentive for one side to strive for a technological breakthrough would be enhanced if the other side's force were confined to a single component.

So the single-pronged force, though in one sense more

"limited," might be distinctly more destabilizing to the balance of military forces.

Or consider other questions outside the strategic area, that also require careful study and analysis if progress is to be made. Consider the proposal that has been recently advanced of withdrawing both Soviet and U.S. naval forces from the Mediterranean with backup provided by the Second Fleet in the Atlantic. The Soviet Union now maintains an almost equal number of naval vessels in the Mediterranean, with backup provided from its Black Sea fleet. Whereas the U.S. Mediterranean fleet is built around attack carrier task forces, the Soviet Mediterranean fleet has no attack aircraft carriers, but emphasizes missile cruisers, destroyers, and attack submarines. The Soviet navy relies for its air support on land-based aircraft in the Mediterranean area. This land-based force has grown appreciably in recent years, and is larger than U.S. land-based air forces in the area.[8]

Under these circumstances, would an arrangement in which Soviet naval forces withdrew inside the Bosporus to the Black Sea, and U.S. forces withdrew west of Gibralter, tend to ease tensions and increase prospects for peace in the Middle East? Or would such a move have the reverse effect? Either of these effects could ensue depending on a number of other circumstances:

the further deployment of land-based aircraft by the Soviets or the United States; the expansion or diminution of military assistance to Israel and to the Arab countries; and the pattern of overall relationships between the U.S. and the Soviet Union outside the Middle Eastern area.

All of these examples support one central point. The difficulties and obstacles that lie in the way of arms control are not governed, or in most cases heavily influenced, by the military-industrial complex. The problems are just hard and complex, and the constraints often numerous and severe, quite apart from the putative role of the MIC as an organized pressure group. To say that the MIC prevents or hinders government action toward controlling arms in these fields is both to exaggerate its influence, and to underestimate the fundamental complexities of the problems themselves. Slogans blaming the MIC for our failure to move farther in the direction of arms control not only do the military an unwarranted disservice, but hinder rather than help in finding solutions by diverting attention and effort from the real problem.

At the same time as there are difficult substantive problems unrelated to the MIC, that complicate the achievement of meaningful arms control, there also are circumstances in the international environment that warrant considerable caution in reducing our military strength if damaging consequences for international peace are to be avoided. What are some of these circumstances?

As the United States moves toward accelerated withdrawal of forces in Vietnam, and diminished deployment and basing overseas, what we find in the international arena is an expanding Soviet presence in the Middle East, as well as increasing evidence of such expansion (in bases and in influence) in the Asian area as well. In Congressional testimony, a group of leading British and American students of the Soviet Union emphasized the high likelihood of a more aggressive and expansionist Soviet policy throughout the world, as the USSR's strategic military capabilities reach equality or superiority compared with those of the United States. [9]

Another serious hazard lies in the prospect of nuclear proliferation which does not seem to be receding. Nor can we say that the possibilities of conventional conflict (for example, between North and South Korea), let alone of insurgent wars in the Third World, are negligible. In the light of the Nixon doctrine and for other reasons, the likelihood of our involvement in such conflicts is much less than it formerly was. We are committed, and wisely so, to placing greater alliance on the capacities of friendly and allied countries for their _self_-defense.

And the threshold of provocation that would involve us in direct military support for their efforts is likely to be much higher in the future than it has been. Nevertheless, we should not adopt an ostrich policy if we want to contribute to progress in international peace keeping and stability.

For all of these reasons, military considerations and substantial military capabilities remain important ingredients of responsible policy and conduct by the United States.

In summary, the main obstacles to arms control seem to me twofold: the difficulty and complexity of specific arms control issues and opportunities themselves (not insuperable, but difficult and complex and requiring serious study and analysis); and the realities of the international arena, including the growth of Soviet military capabilities in general purpose as well as strategic forces. Neither of these real impediments to arms control is illuminated or indeed significantly influenced by the imagery and rhetoric accompanying cliches about the "military-industrial complex."

I recently had an opportunity to discuss certain general problems pertaining to arms control with a Soviet official who expressed concern about the power and influence of the industrial part of the U.S. military-industrial complex. He acknowledged that the military establishment in the Soviet Union, like that in the United States, does present opposition to arms control, but felt that this opposition is similar in the two systems. The contrast, he said, lies in the opposition that is provided by the industrial part of the U.S. MIC, which does not correspond to any such opposition in the Soviet Union. As far as U.S. defense industry is concerned, he went on, it differs from that of the Soviet Union because of the profits that it might stand to lose, and because it is not subject to government discipline and control. On the other hand, in the Soviet Union, he asserted, defense industry is secure in knowing that it will be fully employed in any event, and so does not have the powerful incentive to persist in military production once the government decides otherwise.

Since variants of this view are current in the United States as well, it is well worth examining. The argument doesn't stand up to critical examination. For example, in the United States, the distinctly limited political potency of defense industry is indicated by a number of glaring facts: over 100,000 engineers and technologists currently unemployed; high rates of general unemployment in the aerospace industry well above that in the nation as a whole; especially sharp reductions in defense budget expenditures for strategic offense and defense forces, as noted earlier; heavy discounting by the stock market of the market prices of defense industry stocks; and two United States

Senators from the State of California, where the aerospace industry is concentrated, who are vociferous opponents of the MIC and advocates of arms control!

On the other hand, if one inquires about similar sorts of indicators of the limited influence of high technology defense industry in the Soviet Union, it is much harder to come by. Instead, what one finds in recent years, is buoyant expansion (prosperity) in these sectors of the Soviet economy, amply fueled by large and growing outlays for research and development, and procurement. In fact, given the profound inertial momentum of large organizations under any system, one may well be skeptical that public sector defense industry in the Soviet Union will be nearly as docile in accepting reduced arms expenditures as private sector defense industry has been in the United States.

In conclusion, I would not want to leave the impression that the military establishment and industrial firms in the United States are likely to be other than cautious, reserved, and skeptical towards arms control. Further, the previous discussion has already suggested a number of reasons why such caution and skepticism are warranted. But there is a more important point to be stressed. If and as the basis for such caution and skepticism diminishes -- for example, by more careful attention and better answers to the serious issues and problems noted earlier, and by the progress of improved understanding and increased contacts between the Soviets and ourselves -- it seems to me evident that opposition by the MIC to further arms limitations would be minor and ineffectual.

In an effort to keep matters in perspective, I would make a still stronger argument. The MIC is a much <u>weaker</u> source of opposition to lower defense budgets and increased arms limitations than are "other industrial complexes" (OICs)[10] to reduction of costs and increases in efficiency in their own special domains. Let me be specific.

Both education and health-care are "industrial complexes" whose scale is of the same magnitude as that of the entire defense sector. The educational-industrial complex involves annual outlays of about $65 billion, and health care expenditures are somewhat above that figure. If one looks inside the large black box that is the educational-industrial complex, one finds a most impressive and depressing result from the considerable research undertaken in the last ten years: cognitive measures of educational effectiveness are not consistently influenced by increased resource allocations for education. The result seems to apply at primary, secondary, and higher educational levels, and applies whether the increased resource allocations are in the form of teachers, buildings, or technological aids of one sort or another. Evidently, social, family and environmental factors are the main factors affecting cognitive performance, not educational inputs.

An obvious inference follows from this depressing result: some, perhaps substantial, <u>reductions</u> in resource inputs and costs should be possible without any loss of educational outputs at least as concerns the cognitive dimensions of educational performance. And there is no clear evidence to suggest that <u>non</u>-cognitive measures, fuzzy and primitive though they be, are reliably linked with the formal educational process. One source of substantial savings would be to differentiate among teachers' salaries according to their relative market values. Mathematics and science teachers in high school, for example, would get higher pay, and teachers of physical education, home economics, and history, with comparable seniority, would get lower pay.

I dare say that the power of teachers' unions and school administrators to resist this or other means of lowering costs and increasing efficiency in education is vastly greater than the power of the military-industrial complex to resist arms limitations!

Turning to health care, one finds a similarly depressing picture of high costs and limited effectiveness. For example, if one compares the performance of the health industry in Great Britain under its National Health Service, with the health industry in the United States, the picture that emerges is severely prejudicial to the United States. Infant mortality and maternal death rates are appreciably lower in Britain than in the United States; 25 years ago the discrepancy was substantial in the opposite direction, with Britain's rates appreciably above those in the United States. As Anthony Lewis has recently pointed out in a striking series of articles in the New York Times, if one looks at seven leading causes of death in the United States and Great Britain over the past two-and-a-half decades (respiratory tuberculosis, diabetes, arteriosclerotic disease, heart disease including coronary, hypertensive heart disease, influenza, pneumonia and bronchitis), in every case the death rate has risen less, or fallen farther, in Great Britain than in the United States.[11]

At the same time, the costs of health care in the United States have been rising much faster than those in Great Britain. Currently nearly 7 per cent of the GNP is expended on health care in the United States; in Great Britain, the figure is less than 5 per cent. How might costs be lowered and efficiency improved in the health industry? One possibility surely lies in substitution of paramedical for medical personnel, an innovation which, incidentally, has progressed much more rapidly in military, than in civilian, medical care. In this connection, it is worth noting that Horton's recent study of medical productivity in the Navy showed a much higher rate of increase (measured in terms of patient care per physician day) than that which occurred in the private medical economy as a whole.[12] Another possibility lies in allowing the capacity of existing medical schools for producing doctors to expand more rapidly, as well as to allow greater freedom of entry into the field.

Once again, opposition by the health-industrial complex -- in this case, the American Medical Association, the County Medical Associations throughout the country, the American Hospital Association, and so on -- is a much more potent obstacle to innovation and change, than is the MIC in relation to arms limitation.

The point is not just tu quoque. When we talk about the MIC, arms limitations, and obstacles to arms control, it is important to maintain perspective and balance if simplicities are to be avoided, and realities appreciated. A reminder about the "other-industrial complexes" can contribute to such perspective, as well as to less of the shrill rhetoric that abounds these days.

NOTES

[1] Sidney Lens, The Military-Industrial Complex (Philadelphia and Kansas City: Pilgrim Press, 1970). All quotations are from pp. 99-100, underscoring added. Other books of the same genre include Richard J. Barnet, The Economy of Death (New York: Atheneum, 1969), and Erwin Knoll and J. N. McFadden (eds.), American Militarism 1970 (New York: Viking Press, 1969).

[2] Adam Yarmolinsky, The Military Establishment (New York: Harper and Row, 1971), p. 6.

[3] New York Times, July 15, 1971, p. 6, and Time, July 26, 1971.

[4] I am indebted for these computations to Timothy V. Wolf. The computations are based on pooling House and Senate votes for each State delegation on each of the three issues. The legislation on the ABM SAFEGUARD system was S2546 and HR14000 (Congressional Quarterly, August 8, 1969 and October 10, 1969); for Cooper-Church, HR15628 (Congressional Quarterly, July 17, 1970 and January 15, 1971); and for the SST, HR17755 (Congressional Quarterly, December 11, 1970 and January 15, 1971). The figures on defense expenditures per capita by State are taken from the Statistical Abstract of the United States for 1969.

The significance of the rank order correlation coefficient, r_s, was tested by the value of t, where

$$t = r_s \sqrt{(n-2)/(1-r_s^2)}$$, with n-2 degrees of freedom, and n = 50.

[5] Defense Industry Profits Study, Report to the Congress by the Comptroller General of the United States (Washington, D.C.: Government Printing Office, March 17, 1971).

[6] Ibid., pp. 66 and 67.

[7] In characterizing these models, I am indebted to analytical work underway by my colleague, John Despres. I have also drawn from an earlier paper of mine, "Military-Industrial Complexities," Science and Public Affairs, Bulletin of the Atomic Scientists, XXVII (February, 1971), pp. 19-22.

[8] See The Military Balance, 1970-71 (London: Institute for Strategic Studies, 1971); and Lawrence L. Whetten, The Soviet Presence in the Eastern Mediterranean (New York: National Strategy Information Center, 1971).

[9] Hearings Before the Subcommittee on National Security and International Operations, Committee on Government Operations, United States Senate, December 1969 - May 1970, (Washington, D.C.: Government Printing Office, 1970).

[10] See "Military-Industrial Complexities," Science and Public Affairs, Bulletin of the Atomic Scientists, op. cit., pp. 20-22.

[11] The New York Times, Oct. 2, 1971.

[12] An Economic Analysis of Progress in the Medical Care of the U.S. Navy and Marine Corps Personnel (unpublished Ph.D. dissertation, University of Washington, 1966).

AN EMPIRICAL STUDY OF MILITARY-INDUSTRIAL LINKAGES

Stanley Lieberson

A major controversy exists in macrosociology over the forces which influence national policy in the United States and other advanced industrial societies. On the one hand, the elitists argue that a relatively small group of people, representing a narrow range of interests, determine national policy across a range of domains including foreign affairs, military spending, and major domestic programs. By contrast, the pluralist school views national decision making as a process influenced by a broad and diverse array of interest groups, with no single group or set of groups powerful enough to consistently dominate the national political system. Although the two schools are not diametrically opposed, as Kornhauser's comparison (1968) suggests, they are contradictory on a number of counts.[2] However, despite the fact that national political systems are a central concern in sociology, efforts to untangle the conflicting theories often appear more like ideological debates than contributions to scientific knowledge.

There are two special obstacles to overcome before the contradictory implications of elitism and pluralism can be resolved. First, scholarly positions on this subject must be judged by their ability to develop a comprehensive system that takes account of the available information rather than by their implications for contemporary politics. Because of the political overtones, both the elitists and pluralists approach the issues polemically, as if total rejection of the rival theory is the only possible way such a dispute can be resolved. It is true that some scientific controversies are eventually resolved when a body of empirical research persistently supports one theory and is incompatible with another.[3] But rival theories in the social sciences often turn out to be incomplete and distorted parts of a greater truth, such that neither is consistently supported by empirical research. The theoretical controversy over heredity versus environment, for example, is now rejected because it involved a false and unnecessary polarization.

The second major obstacle stems from the difficulties of conducting empirical research relevant to these theories. Not only are sociologists unable to make direct observations on the decision processes, but it is difficult to design research that encompasses the broad range of political events covered by these conflicting theories. No single research effort can be sufficiently comprehensive or critical, for example, to provide a conclusive test of the Marxist argument that capitalism generates a surplus that can only be used up through war and waste (see Baran 1957; Baran and Sweezy 1966). As a consequence, overinterpretation of the results is all too easy. Actually, each empirical study must be viewed only as contributing to an aggregate of investigations that deal with national power.

MILITARY-INDUSTRIAL LINKAGES

This study focuses on one facet of the theoretical controversy, the relation between military expenditures and large corporations in the United States, the so-called military-industrial complex. According to one leading elitist (Mills 1959, p. 276), "American capitalism is now in considerable part a military capitalism, and the most important relation of the big corporation to the state rests on the coincidence of interests between military and corporate needs." Mills goes on to assert that there is a "coincidence of interest between those who control the major means of production and those who control the newly enlarged means of violence." Pluralists also recognize the existence of a military-industrial complex (see, e.g., Rose 1967, pp. 94-98), but they view this as but one of many interest constellations. Accordingly, they deny the elitist claim that this particular complex has unique breadth, scope, and coordination (see the review by Pilisuk and Hayden 1968).

Several facets of the dispute can be submitted to empirical study. Indeed, unless one claims that power and influence are unrelated to position in the social structure, it is clear that many of the issues raised by the elitists and pluralists call for a quantitative approach, dealing with matters of degree and frequency rather than absolutes. How extensive are the overlapping interests of major institutions? To what degree, and in what domains of national policy, do these connections modify and influence the decision-making processes? Has there been a substantial change in these relations over time?

There is little reason to doubt that a "military-industrial complex" exists if by this phrase is meant a set of commonly shared interests between the military and some major corporations. Certain striking features of these linkages, recognized by both theoretical schools, are: the interchange of personnel between the military and their corporate suppliers, the network that exists within the business sector, and the role of large corporations as suppliers to the military.

Circulation of Personnel

The 100 largest primary military contractors in the 1957-58 fiscal year, recipients of three-quarters of the money awarded, employed 218 former generals or admirals (Janowitz 1960, p. 376). Altogether, these 100 contractors employed some 768 former military officers who had retired with at least the rank of colonel or naval captain. The linkages have increased several fold during the past decade. The 100 largest primary military contractors in the 1968 fiscal year employed 2,072 former military officers who had been colonels, naval captains, or higher.[4] Excluding five companies for which data are not available, this means an average of twenty-two former officers per major contractor.

The increasing interlock between suppliers and the military during the decade suggests the possibility of a growing community of interest between the two sectors of the society. Personal contacts between high-ranking officers and their former colleagues can affect negotiations, particularly when the military officers themselves may soon seek employment from the corporations after military retirement (U.S., Congress, Senate 1969, p. S3072-3). Admiral Hyman Rickover has described a number of ways that military contracts may be negotiated to maximize the contractor's gain, for example, contract manipulation, research and development advantages, patents, shoddy procedures for cost accounting, and the like (Joint Economic Committee 1968).

Coordination Among Corporations

Underlying the power-elite thesis is the proposition that large corporations operate in concert, coordinating their activities and interests. "Would it not be strange," Mills (1959, p. 123) asks about large businesses, "if they did not consolidate themselves, but merely drifted along, doing the best they could, merely responding to day-to-day attacks upon them?" One indicator of this elite community is the degree of interlocking directorships among large corporations. "As a minimum inference, it must be said that such arrangements permit an interchange of views in a convenient and more or less formal way among those who share the interests of the corporate rich."

Interlocks among boards of directors is a major topic for research in itself. However, an illustration based on leading banks does give the reader a perspective on the kind of linkages that do occur and the economic resources involved. Listed in table 1 is the board composition of the fifth-largest New York City bank in early 1965. The board represents leading corporations from many sectors of the national economy, creating the potential for a rather extensive and diverse communication network. In parentheses, alongside the name of each corporation, is its national rank. The sweep of interlocks is understated, actually, since only the primary affiliations of outside members of the bank board are shown.

Altogether, the boards of the seven largest New York City banks in 1965 include officials from fifty-one of the largest 500 industrial companies. There is a particularly heavy concentration from the fifty largest companies, with nineteen represented on the boards of at least one of the seven banks. The boards also have officers from some of the largest transportation, merchandising, and life insurance companies.[5] Total employment in 1964 among the companies represented on these bank boards range from 400,000 to 1.65 million, with the median bank board representing the employers of 1.5 million.[6]

Major Contractors

Clearly a segment of American industry is deeply dependent on the military for survival. Attempts by the aircraft industry, for example, to diversify and reduce their dependency on the government have been unsuccessful for the most part (Weidenbaum 1963, pp. 79-83). Even more significant, major industrial corporations obtain a lion's share of the largest primary military contracts. Three-fifths of the fifty largest industrial corporations (as measured by sales in 1967) are among the 100 largest military contractors for the 1968 fiscal year (see table 2).

Altogether, sixty-eight of the 100 biggest contractors were among the 500 largest industrial corporations in the United States. Moreover, the remaining thirty-two contractors include two leading utilities (ranking first and forty-third in assets among U.S. utilities) and four leading transportation companies (ranking among the leading fifty in operating revenues). In short, three-fourths of the largest military contractors are major corporations.

Implications

The results reported above, like those in a number of earlier studies, are suggestive of a close tie-up between the military and at least some industrial companies. But these reports are peripheral to an evaluation of the two theoretical approaches. For pluralists and elitists do not differ on the question of whether there is a close tie-up between some industrial companies and the military, rather they differ on the <u>magnitude</u> of the interlock and its <u>causes</u>. In order to evaluate these approaches, it is necessary to examine the relative importance of military spending for large businesses. Pluralists view military-industrial relations as but one of many powerful influences on government policy; whereas the elitists see this as a dominant and pervasive influence caused by an inherent necessity for the survival of a capitalistic system.

TABLE 1

Outside Board Members of the Morgan Guaranty Trust Company, New York, 1965, by Primary Company Affiliation

Corporation (and Rank)	Principal Activities
American Machine and Foundry (150)	Bowling, leisure time products, specialty industrial machinery
American Telephone and Telegraph (U-1)	"Bell" system, electronics
Bethlehem Steel (18)	Steel
Campbell Soup (93)	Soup, other food products
Coca Cola (68)	Soft drink syrup, juices
Columbia Gas System (U-10)	Natural gas distributor
Continental Oil (39)	Crude oil, fertilizer, petrochemicals
E. I. duPont de Nemours (12)	Chemicals, synthetic fibers
Gillette (204)	Razors, wave kits, pens
International Nickel (F-70)	Nickel, platinum, copper
New York Life Insurance (I-4)	Insurance
Pennsylvania Railroad (T-2)	Railroad
Procter and Gamble (24)	Soap, foodstuffs from vegetable oils
Singer (60)	Sewing machines, calculators
Standard Oil of New Jersey (2)	Petroleum
J. P. Stevens (86)	Textiles
State Street Investment	Mutual fund

Note.—Data not available on two additional organizations represented on the bank's board: Bechtel Corporation; The Duke Endowment. Letters F, I, T, and U indicate corporation is ranked, respectively, as foreign, life insurance, transportation, or utility. All other rankings refer to 500 largest industrial corporations.

Source.—Rankings from Fortune (1965a, 1965b); principal activities from Standard and Poor's (1965).

TABLE 2

Overlap between 100 Largest Primary Military Contractors and 500 Largest Industrial Corporations, 1968

Rank of Industrial Corporation	Number among 100 Largest Contractors
1–50	29
51–100	11
101–150	5
151–200	7
201–250	3
251–300	4
301–350	5
351–400	1
401–450	1
451–500	2
Other*	6
Total	74

Source.—One hundred largest military contractors in 1967–68 fiscal year from Congressional Record (U.S., Congress, Senate 1969). Largest 500 industrial corporations and other leading nonindustrial corporations obtained from Fortune (1968).

* Includes nonindustrial companies that are among the fifty largest commercial banks, life insurance companies, merchandising firms, transportation companies, and utilities.

DEPENDENCE OF INDUSTRY ON A WAR ECONOMY

Military Contracts

As indicated in table 2, the very largest corporations obtain most of the major military contracts. However, the issue now is the importance of these contracts for such corporations. Both Weidenbaum (1965, pp. 113-14) and Lapp (1968, pp. 186-87) have demonstrated that the ratio of military contracts to total sales varies widely among leading contractors. In order to obtain some estimate of the role of military expenditures for large businesses generally, it is more appropriate to focus on leading industrial companies in the nation rather than on simply the leading contractors.

Among the fifty largest industrial companies, all with sales well over $1 billion in 1967, primary military contracts vary greatly in significance (see table 3). Contracts obtained by General Dynamics, the largest military contractor in the 1968 fiscal year, were nearly equal to its total sales in 1967 (the ratio is .993). Another corporation has a contract-sales ratio of more than .75; one has a ratio of .60; and there are seven others with ratios exceeding .25. Twelve of the fifty largest industrial companies, on the other hand, have ratios of .04 or less -- even though they are among the 100 largest military contractors. Military ties are very minor for an additional

TABLE 3

RATIO OF MILITARY CONTRACTS TO TOTAL SALES AMONG 100 LARGEST INDUSTRIAL CORPORATIONS, 1968

RATIO	NUMBER	
	50 Largest Corporations	50 Next Largest
.75 and above	2	0
.50–.74	1	2
.25–.49	7	4
.10–.24	1	5
.05–.09	6	24*
Under .05	33	15

SOURCE.—See table 2.

* Ratios for all twenty-four of these companies are estimated in terms of maximum possible values. Undoubtedly, most would have lower ratios if actual contract data were available (see text).

TABLE 4

REGRESSION ANALYSIS: CORPORATE INCOME AS A FUNCTION OF FEDERAL GOVERNMENT EXPENDITURES, 1916–65

Variable	r	b
YX_1	.68	.30
YX_2	.93	.77
X_1X_2	.76	1.42
$YX_1 \cdot X_2$	−.08	−.02
$YX_2 \cdot X_1$	.82	.89

SOURCE.—X_1 based on U.S. Bureau of the Census (1960; 1965a, column Y358; 1965b, table 534); X_2 based on U.S. Bureau of the Census (1960; 1965a, column Y357 minus column Y358); Y based on U.S. Bureau of the Census (1960; 1965a, column Y283 minus columns Y288 + Y289); and 1962–1965 data based on U.S. Bureau of the Census (1968, table 694).

NOTE.—X_1 = federal government expenditures for national security (in millions of dollars); X_2 = federal government expenditures not for national security (in millions of dollars); Y = net corporate income after taxes (in thousands of dollars).

twenty-one companies that are not listed among the top 100 contractors. Even if we make the extreme assumption that the contracts received by these companies are only $1,000 less in value than the one-hundredth-largest contractor, the ratio of military to total sales would be between .03 and .04 for eleven companies and .02 or less for ten companies. Altogether, then, thirty-three of the fifty largest industrial corporations have contract-sales ratios of .04 or less.

Among the fifty next largest industrial corporations, there are thirty-nine that are not on the list of 100 leading military contractors in 1968. Again making the extreme assumption that each of these holds contracts that are only $1,000 less than the one-hundredth-largest military contractor, the ratios would be .04, .05, and .06 for, respectively, fifteen, nineteen, and five of these corporations (table 3).

It is clear that the majority of the largest industrial corporations derive only a small portion of their total business from primary military contracts. Moreover, there are numerous merchandising companies that do not enjoy any direct benefits from military contracts. Among these companies, there are twenty-one with sales equal to at least the one-hundredth-largest industrial company (eight have sales greater than the fiftieth-largest industrial company). Undoubtedly, stores are located in communities that receive sizable military contracts, but military expenditures per se, as opposed to government spending in other sectors, is probably not particularly beneficial to major retailers.

Although the data reported above fail to support the notion that American industry is deeply dependent on a military economy, the results are hardly conclusive. First, they fail to take into account the indirect consequences of military spending for the nation's major corporations. Large corporations may supply many of the primary contractors and thus benefit from the demands created by military expenditures. Steel mills, for example, produce the armor plate used in tanks or needed for the hulls of war vessels. Further, there may be general indirect benefits that are based on the prosperity generated by military expenditures. If military spending pumps large amounts of money into the economy, then consumer and industrial demands may be of substantial benefit to all sectors of the economy, not merely the "munitions makers." Finally, military contracts may still be very important if they are unusually profitable or utilize plants and equipment that might otherwise be idle. For General Motors, the largest industrial corporation in the nation and the tenth-largest military contractor, the ratio of military contracts to total sales is only .031.

However, the contracts do amount to nearly two-thirds of a billion dollars, a sum that can hardly be considered trivial.

In short, the data fail to support the hypothesis that large American businesses are deeply dependent on military spending. The results are inconclusive, however, because there may be substantial indirect benefits for a broad spectrum of American businesses that are not military contractors.

Regression Analysis

Regression analysis provides another method for estimating both the absolute and relative influence of military expenditures on <u>total</u> corporate income. The dependent variable, Y, is corporate income after taxes for each year between 1916 and 1965. Two independent variables are used: government expenditures in each year for "major national security" (X_1); government expenditures that are not for national security (X_2). Although the latter includes veterans benefits and other costs reflecting earlier military efforts as well as space explorations, X_1 essentially measures current military expenditures over which the government has some option, and X_2 reflects the nonmilitary expenditures of the government. Overall, there is a high association between corporate income and the two facets of government expenditures; the coefficient of multiple determination $R^2 y \cdot x_1 x_2$, is .88. This is not altogether surprising, given inflationary trends through the period as well as the necessary relationships between corporate income and government expenditures. On the zero-order level, corporate income is more closely linked to nonmilitary expenditures than to military spending (see table 4). Even more significant, military expenditures have less impact on corporate income than does an equivalent amount spent by the government on nonmilitary items (compare b_{yx_1} with b_{yx_2}). Both the correlation and regression of corporate income on military expenditures are virtually nil after nonmilitary spending is taken into account. By contrast, the partial correlation and regression for income on nonmilitary expenditures remain very high, .82 and .89, respectively.

Following the procedures described by Blalock (1961, 1964) and Fendrich (1967), the zero-order and partial correlations fit two models. The first one shown below is supportive of Mills, whereas the second model is not. Both models suggest that: $r_{yx_1 \cdot x_2}$ be approximately zero; that r_{yx_1} be lower than either r_{yx_2} or $r_{x_1 x_2}$; and that r_{yx_1} be approximately equal to the cross product of the other two correlations. All of these conditions are met. Without additional variables, it is impossible to distinguish between them.

$$X_1 \longrightarrow X_2 \longrightarrow Y$$

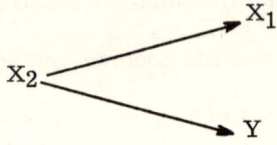

Even under the first causal model, however, efforts to increase corporate income would be directed towards raising nonmilitary spending rather than the military budget. First, it is necessary to recognize that the two facets of government spending operate somewhat independently of each other (r^2 is .58). This means that there are other factors influencing nonmilitary spending. Since the regression coefficients indicate that a unit change in nonmilitary spending will increase corporate income far more than will an equivalent military expenditure, the first model would still suggest that nonmilitary spending receive preference by those seeking to raise corporate income. In short, during the past fifty years a dollar spent on the military appears to have generated far less corporate income than a dollar spent by the government in other realms.

Although the results fail to support the contention that corporate income is deeply dependent on military expenditures, again the analysis is not conclusive. For one, the relationship between government spending and corporate income involves some feedback, a complexity not found in the models reported above. In addition, other lag combinations could be used besides the one where corporate income trails government expenditures by a half year. Moreover, it is entirely possible that the political limits on nonmilitary spending are much greater than for military spending. If a shift from the military budget to nonmilitary needs interferes with private enterprise, then the military option might be encouraged. Finally, elitists can argue that military spending provides an umbrella for foreign investments. Such expenditures are therefore necessary, the argument would run, even if a greater immediate return is possible from a budgetary shift to nonmilitary spending.

Input-Output Analysis

An additional method for determining both the indirect and direct consequences of military spending for various sectors of the economy is provided by input-output economics. For each specific industry, this quantitative tool in economics considers the role of every supplier and every market, yielding an interindustry matrix of economic supply and demand. Dealing with the production of automobiles, for example, this method indicates the amount of rubber, steel, glass, copper, aluminum, etc., required and then, in turn, the supplies needed by the producers of these products, say sulfur for rubber, and so on.

Leontief and his associates (1966, pp. 184-222) investigated the role of military expenditures through this perspective. A

20 per cent cut in armament expenditures accompanied by a compensating increase in nonmilitary expenditures would mean a reduction in total output and employment in only ten of the nation's fifty-six industrial sectors (Leontief et al. 1966, pp. 194-97). Among the ten industries with declines, five would suffer percentage losses exceeding the percentage gain enjoyed by any single one of the remaining industries. These five industries are: aircraft (-16 per cent), ordinance (-15 per cent) research and development (-13 per cent), electronics equipment (-5 per cent), and nonferrous metals (-2.2 per cent). The sector with the greatest increase, agricultural services, stands to gain only 2.1 per cent.

These results, although basically an exercise in economics, have profound sociological implications for the pluralism-elitism controversy. They mean that the economic consequences of a substantial step toward disarmament are lopsided; most industries (particularly the first three named above) would be far greater than the percentage gained in any single sector. Although the economy as a whole is not harmed by such military cutbacks, the small number of industries with a substantial vested interest in military spending would suffer far more than other industries would gain. This means that the industries benefiting from military expenditures are more narrowly concentrated than are the economic interests that stand to gain through disarmament. The implications of these conclusions are considered in greater detail later.

An even more elaborate application of input-output analysis by Kokat (1967, pp. 805-19), based on the assumption of a 50 per cent cut in the defense budget, also leads to the conclusion that a shift to nondefense expenditures would have an expansionary impact on the majority of industries. If there are compensating increases within either the private or public sectors, only a very limited number of industries would suffer from such a severe cutback in military expenditures.

Although input-output analysis fails to support the hypothesis that military spending is a prerequisite to corporate prosperity, an alternative interpretation is possible that is consistent with elitist theory. Despite the results reported above, an elitist might argue, shift to peacetime consumption would be resisted if the transitional costs were very high in compairson with the gains that might ensue. Retooling, new marketing procedures, investment in specialized plants and equipment, greater competition, and other costs might well outweigh the small benefits in sales and employment that most sectors of industry would enjoy through a shift from military to nonmilitary expenditures. Moreover, an abrupt cutback in military spending would have serious social and economic consequences in a number of communities. Large cities such as Fort Worth and Seattle, as well as some industries, would be particularly affected by drastic changes in military policies (Halverson 1969; Ames 1969).

HISTORICAL TRENDS

Rather than restrict the issue to elitism versus pluralism, at this point one could argue that neither theory provides an adequate interpretation of the available data. Although reinterpretation consistent with the elitist perspective is possible, the data above suggest that a high level of military spending is not vital to the prosperity of either the nation or its largest businesses. On the other hand, given the findings of Russett (1969a, 1969b) that a wide variety of consumer and other public civilian activities suffer when there are substantial military expenditures, the pluralist perspective is not entirely satisfactory. For, if more interest groups would benefit from alternative government-spending policies, then how can pluralist theory explain the maintenance of such a substantial military commitment?

Since it is almost inevitable that any valid macrosocietal theory have implications for social change, a historical perspective provides additional clues to the theoretical issues at hand. The historical trends for both government spending, generally,

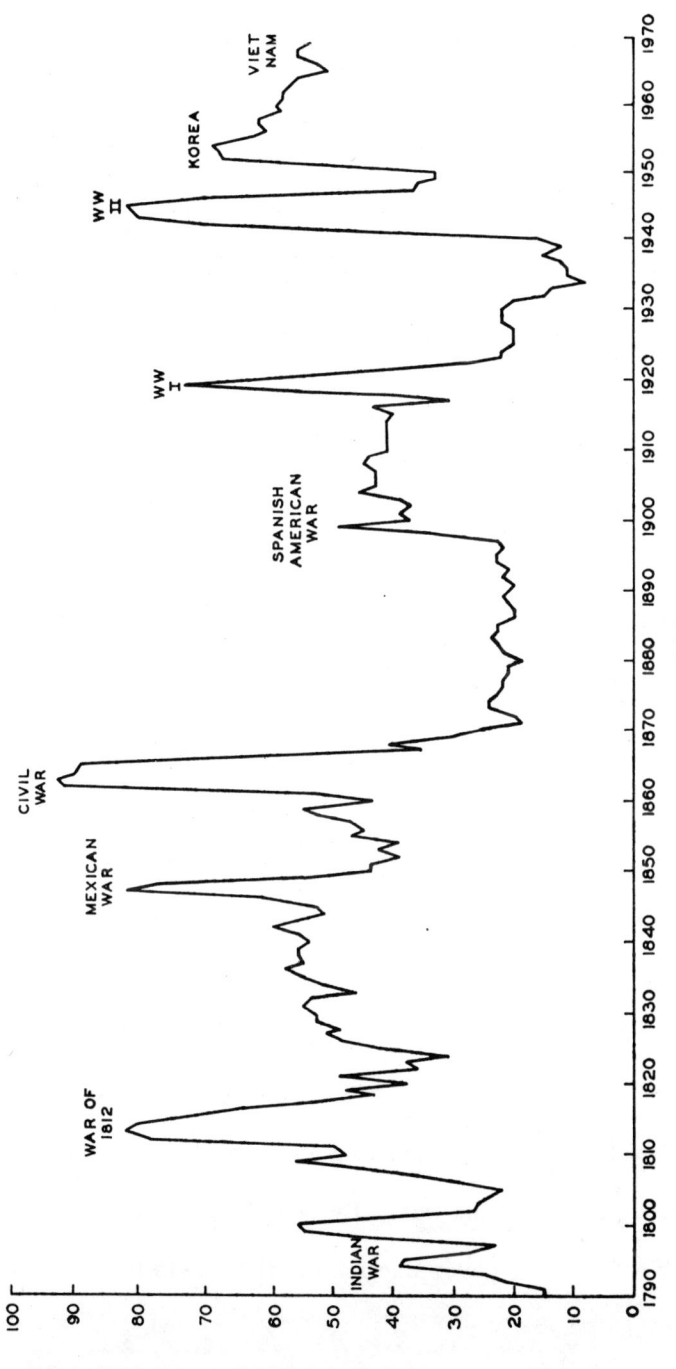

FIG. 1.—Percentage of the federal budget for the military, 1790–1969. Sources: U.S. Bureau of the Census (1960, columns Y350 and Y358; 1965a, columns Y350 and Y358a; 1965b, table 534; 1968, tables 538 and 544), based on expenditures for the "armed services" before 1900 and for "major national security" or "national defense" after 1900. There are some minor inconsistencies among sources.

and military spending, in particular, fail to fit neatly into either the pluralistic or the elitist interpretations.

The percentage of federal expenditures devoted to the armed forces has fluctuated considerably during the nation's history, ranging from about 10 per cent during the Great Depression to over 90 per cent during the Civil War (fig. 1). There is no evidence that military spending currently occupies an unusually large proportion of the total federal budget; rates during the 1960s are no higher than those in the 1830s and 1840s. The United States, compared with other nations, spends a relatively large part of its gross national product (GNP) on the military (see Benoit and Boulding 1963, pp. 301-6), but the current percentage is by no means uniquely high when compared with the nation's past record. In this regard, there is no support for the thesis that the role of the military in the government has expanded in recent decades (Mills 1959).

On the other hand, there is some evidence that military spending after the Second World War failed to drop as sharply as it had after previous wars. The War of 1812, the Mexican War, the Civil War, and World War I were followed by periods of very low levels of military spending -- even when compared with the period preceding the war. By contrast, military expenditures after World War II never declined to the levels found in the

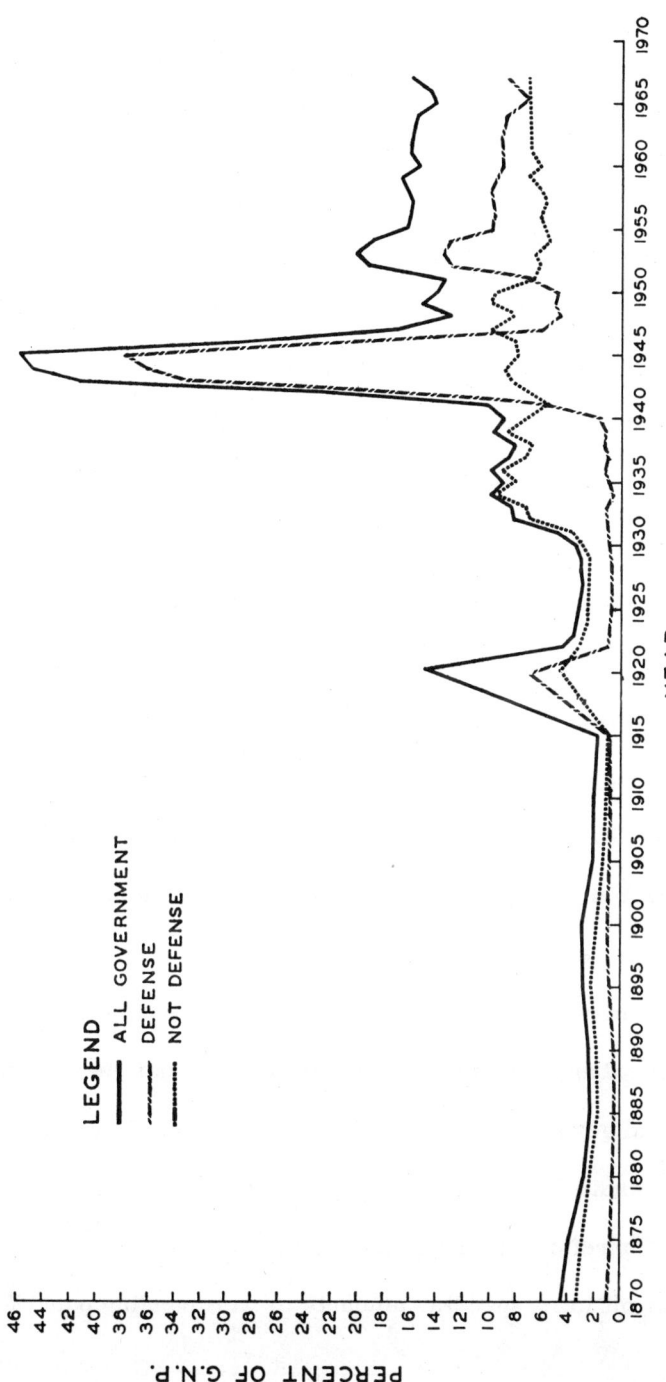

FIG. 2.—Expenditures of the federal government as a percentage of the GNP, 1869–1967. Sources: In addition to those listed under fig. 1 U.S. Bureau of the Census (1960; 1965a, column F1; 1968, table 454).

1930s. This feature is compatible with the elitist perspective which emphasizes the change in military-industrial relations that occurred in recent decades.

Although the proportion of government expenditures devoted to the military shows no clear-cut temporal trend, a historical analysis does reveal that the role of government in the national economy has consistently increased through the years. From the postbellum period until the First World War, government expenditures ranged between 2 and 3 per cent of the GNP (fig. 2). After a sharp increase during World War I, the percentage remained above earlier peacetime levels. The federal budget increased to 10 per cent of the GNP during the 1930s and reached even higher levels after the emergencies created by World War II had ended. Current government expenditures are about 15 per cent of the GNP.

Figures 1 and 2 suggest that changes in the role of the military in the national economy are to be seen largely as a consequence of changes in the role of government in the economy. Consumption of an increasing portion of the total national output by the federal government means that both its nonmilitary and military expenditures are of growing significance to the total economy. Therefore, although the elitists are correct in arguing that military spending amounts to an expanding part of national productivity, the fact remains that the federal government's impact on many other sectors of the society is also growing. In the case of research, for example, the government plays an exceedingly important role (Barber 1968, p. 225). Likewise, there is some evidence of an "education-industrial complex" that operates with considerable effectiveness in influencing federal expenditures for education (Miller 1970). The government's ability to command an increasing portion of national productivity through taxes and other revenues means that a focus solely on the changing military-industrial relations could lead to a very misleading conclusion.

In short, since this brief review of historical trends fails to consistently support either the elitist or the pluralist approaches, it appears all the more reasonable to consider another perspective. Suggested below is an alternative interpretation of military-industrial relations based on what might be called the mechanism of compensating strategy.

AN ALTERNATE PERSPECTIVE

A high level of military expenditures is not necessarily due to a broad set of intense vested interests in such a policy. To be sure, there are both industries and entire communities that are deeply dependent on military expenditures. Moreover, once a pattern of military consumption is established, a shift to non-military spending would be costly for some industries and regions that were not initially dependent on military spending. Nevertheless, when compared with the consequences of alternate government policies, most sectors of American industry gain very little from a high level of military expenditures.

On the other hand, the absence of widely shared direct benefits from military spending need not be taken as evidence that a small segment of American industry is dominant. Given the different economic and organizational resources that various segments possess, obviously interest groups in the society do not have equal power. Compare, for example, the resources available to the urban poor with those available to the petroleum industry. Nevertheless, policies detrimental to the numerical majority will occur even if a smaller group does not dominate the entire political system. Rather, a process of decision making may be postulated in which the political system's output can fail to reflect the interests of the majority without this necessarily meaning that a power elite controls the government. Even if interest groups were alike in organization, wealth, and other competitive resources, under certain conditions a "compensating" strategy could be pursued such that political decisions would not invariably reflect the direct interests of the majority.

In order to specify the conditions necessary for a compensating power system to operate, two additional characteristics of interest groups should be noted. First, populations within advanced industrial societies are atomized into a highly diverse set of groups, with interests that are neither fully harmonious with one another nor fully competitive. Not only does an expanding division of labor subdivide a nation into increasingly specialized needs and vested interests, as Durkheim noted, but there is also an increased territorial division of labor such that spatial subunits of a nation also develop distinctive concerns. A second key element involves recognition of the fact that interest groups differ more than in terms of whether they are for or against a specific proposal. Rather they also differ in the _degree_ of importance that each issue holds. What this means is that groups vary widely in how far they are prepared to go to achieve a specific end.

Accordingly, an advanced industrial society, regardless of its political ideology, must be viewed as one in which a wide variety of interest groups are each attempting to balance an array of potential gains and losses in order to generate the maximum net gain. Advantages that might accrue from opposing another segment are not necessarily pursued if even larger gains are possible through the expenditure of an equivalent amount of political capital in some other government domain. In other words, if interest groups are not by themselves able to obtain the political outcomes they desire, then it is necessary for them to form alliances and coalitions with other interest groups. Such combinations mean that each interest group sacrifices some issues in order to obtain the greatest net rewards. As a consequence, a political body's decision with respect to a particular issue may reflect the intense concern of a minority of interests coupled with the support obtained from other segments whose major interests are found elsewhere.

Four conditions are necessary for a "compensating" power situation to operate. First, power must be an exhaustible commodity such that each interest group has limited political influence. Second, the interests of the population must be diverse enough so that a given proposal will not have an equal impact on all segments. Third, governmental actions that favor a particular interest group must not eliminate the disadvantaged majority's potential for other gains. In the case at hand, it means that funds are available for more than the military budget or that nonbudgetary legislation can be passed that will yield sufficient gains for other segments of the population. Fourth, legislation beneficial to a specific interest must not at the same time create too great a loss for the majority of other interests. Otherwise, a concerted effort to combat the legislation will provide a net gain to the majority that is even greater than would occur under their use of a compensating strategy.

When these conditions are met, it follows that a given group will not attempt to exert its influence on all issues to the same degree, but rather will concentrate on those that generate the greatest net gain. To obtain passage of particularly beneficial legislation, an interest group may in turn support other proposals that, by themselves, create small losses (see Coleman 1964). The third and fourth conditions are suggested by the controversy over military spending during the Vietnam war. The "guns and butter" policy first meant that other sectors of the population could pursue compensating strategies, such as required by the third condition. On the other hand, the monetary and domestic crises generated by this series of compensating gains, coupled with the unpopularity of the war, make it increasingly hard to meet the last condition. That is, the potential net gains that would be achieved by the vast majority of the population through a decline in military spending begins to approach or exceed the gains that many segments obtain through a compensating strategy.

According to the notion of compensating strategies, military spending can be explained by a high level of interest on the part of one segment of American business accompanied by a relative lack of concern on the part of other segments of industry. The majority of industries disregard the small losses they incur from defense spending and turn their attention to other aspects of government policy that affect them more directly. Defense spending will influence the profits of dairy farmers, for example, but so too will other government policies such as price supports, marketing restrictions, exports and imports on dairy products, grading practices, trucking costs, and so forth. Accordingly, the area of government action with the most substantial direct impact on the goals of the dairy industry is probably not the military.

Indeed, barring a major war emergency, there are other facets of federal activity that are far more crucial to the interests of many industrial segments. The relevance of the federal government is hardly restricted to the military budget; decisions affecting taxes, antitrust laws, foreign markets, imports, minimum-wage laws, and a wide variety of other government regulations are extremely important to business. These facets of government policy often have far greater short-term consequences for a specific industry than do military expenditures. The tobacco companies, for example, stand to lose far more from restrictions on cigarette advertising than they are likely to gain

TABLE 5

Vested Interests of States and the Membership of Senate Committees, 91st Congress, 1969–1970

Committee	All States		States on the Committee	
	Primary Interest*	Secondary Interest	Primary Interest*	Secondary Interest
Agriculture and Forestry	30	20	13	0
Indian Affairs	17	33	10	0
Minerals, Materials, and Fuels	20	30	8	1
Merchant Marine	30	20	9	1
Housing and Urban Affairs	20	30	7	8
Labor	15	33	7	5
Armed Services	25	23	14	3
International Organization and Disarmament	23	25	5	2

Note.—Committee membership based on *Congressional Index* (no date).

* Primary interests of the states in various committees based on the following criteria: *Agriculture and Minerals*—states whose labor force in 1960 exceeds the United States percentage in the agriculture and mining industries, respectively; *Indian Affairs*—states in which the Indian percentage of the population exceeds that for the total United States; *Merchant Marine*—states contiguous to an ocean, the Gulf, or the Great Lakes; *Housing and Urban Affairs*—states exceeding the national percentage of residents who reside in standard metropolitan statistical areas; *Labor*—states exceeding the national percentage of population in the AFL-CIO (data available for forty-eight states); *Armed Services*—states that would suffer a net loss in output and employment after compensated cut in armament expenditures (Leontief et al. 1966, p. 197) (data available for forty-eight states); *International Organization*—same as above, except states gaining in output and employment (data available for forty-eight states).

in the event of disarmament.[7] The petroleum industry would also enjoy a small gain if there was a compensated cutback in military expenditures; however, federal budgetary decisions with regard to their special depletion tax advantages are far more significant. The largest gain would be enjoyed by agricultural services, but obviously agriculture is influenced by many other areas of government legislation and control.

Senate Committees

This propensity for each segment of the society to pursue its areas of greatest direct interest is illustrated by the states represented on various congressional committees. Although residents of all states are in some manner affected by the policies of any committee, it is clear that political efforts emphasize those areas with the greatest potential returns. Hence, each state tends to be represented on committees affecting those facets of the legislative process that are of greatest importance to the state. The House Agriculture Subcommittee on Tobacco, for example, is loaded with legislators from tobacco-producing states; six of the seven members come from Virginia, North Carolina, South Carolina, and Kentucky.

Table 5 illustrates the association between special interests and membership among several Senate committees. A crude index of vested interest is developed for each committee. States ranking above the national average on this indicator are classified as states with a special primary interest in the committee's work; states below the national average are classified as having a secondary interest. The index of vested interest for the Senate Agriculture and Forestry Committee, for example, is the percentage of the 1960 labor force in agricultural and related industries. There are thirty states with a higher percentage than the national figure of 6.7 and twenty states with a lower percentage. These thirty states supply all thirteen members of the Senate's Agriculture Committee.

Although the residents of all states are affected by the prices for food and other agricultural products, the Agriculture Committee is almost entirely in the hands of southern and midwest senators from states with a particularly high vested interest in this domain. Such a pattern is consistent with the proposition that an unequal distribution of potential gains and losses will generate different political thrusts for various interest groups.

Other committees also consist of senators from states with the greatest stake in their activities. The Indian Affairs Subcommitte consists entirely of senators from states with a high proportion of Indians. The Minerals and Fuels Committee is loaded with senators from states with relatively large segments of the labor force engaged in these extractive industries.

Merchant Marine draws senators from coastal states. States with large metropolitan populations are overrepresented on the Housing and Urban Affairs Subcommittee, just as the Labor Subcommittee draws senators from states relatively high in union membership.

In this context, the composition of the Senate Armed Services Committee is not entirely surprising. The committee is loaded with senators from the states that would suffer a net loss if there was a compensated 20 per cent cut in armament expenditures (based on the input-output analysis reported by Leontief et al. 1966, table 10-2). Only three of the seventeen members of this committee are from the twenty-three states that would enjoy a net gain. On the other hand, the small Subcommittee on International Organization and Disarmament Affairs is disproportionately composed of senators from states that stand to gain through a military cutback.

Obviously other forces influence committee assignments in the Senate. Some committees are attractive simply because they may catapult members into the national limelight, for example, Foreign Affairs. Budgetary and procedural committees are extremely powerful, covering a wide range of activities, although it is difficult to link them with specific vested interests. In addition, some assignments are influenced by tradition and seniority. Moreover, this brief examination does not cover the complex exchange and trade networks that may exist between different congressional blocs.

Nevertheless, the results reported above do serve to illustrate the selective pursuit of vested interests. The fact that all activities of the Congress are of consequence in some direct or indirect way to every resident and every organization does not prevent each segment of the society from emphasizing those areas where maximum net gains may be obtained. In this regard, the composition of the Senate's Armed Services Committee may be viewed as but another manifestation of this organizational principle.

DISCUSSION

No single empirical study can determine the nature of military-industrial relations, to say nothing of resolving the elitist-pluralist controversy in which it is enmeshed. However, the data are suggestive both about the empirical question and the theoretical restatement that may be necessary.

In terms of the empirical facets of the study, several elements of the elitist position are supported. Some major corporations are deeply dependent on armament contracts; and the decline after World War II in the proportion of government spending on the military is less than would be expected from the patterns after earlier wars. Moreover, the federal government is a growing factor in the national economy, and hence its expenditures for the military are of increasing importance to the business world. There is no reason to doubt the existence of a military-industrial complex _if_ by that phrase is meant an intense dependency on military contracts among some very large corporations.

On the other hand, there is no evidence to support the contention that the general success of large businesses in the nation depends on substantial expenditures for the military. Indeed, there is evidence to indicate that most would benefit from alternative government spending or from equivalent expenditures in the private sectors of the economy. For the majority of the largest corporations, military contracts at most amount to a very small portion of their total business. Moreover, both the regression and input-output analyses suggest that corporate prosperity would be increased in the absence of military spending.

To be sure, a substantial and sudden drop in military expenditures would create difficulties for a few sectors of the economy and certain cities and regions that are deeply dependent on the military. Nevertheless, the consequences of disarmament should not be interpreted as the cause of its initial buildup. Cities do not first grow and then develop military industries that support them, rather the growth of cities may reflect the expansion of their economic base. Moreover, the bulk of American industry would gain under alternative spending policies.

If these inferences about military-industrial relations are valid, then both of the major current theories provide inadequate perspectives on the nature of societal power. Crucial to the reconceptualization suggested earlier is recognition of the fact that each of the diverse interest groups may, under some circumstances, take stands that are counter to their interests. This will occur when four necessary conditions are met such that a "compensating" strategy is possible. If each group attempts to generate the greatest _net_ gain for itself, then a given policy need not be the product of simply the majority of interests, nor need it mean that a small set of interests is dominant. Rather, the policy may mean that the losses to a majority of interests are small, whereas the gains to some sectors are substantial.

From this perspective, military spending is but one among many sets of vested interests. To be sure, this domain could reach the point where it dominates the entire economic and social life of the nation. However, the results suggest that military expenditures are not, currently, a vital and necessary prerequisite to general corporate prosperity. Using the hypothesis of compensating strategies, the high level of military expenditures need not be interpreted as the product of a common vested interest in such spending among major businesses. Rather, extensive military spending may be viewed as reflecting the operations of an important and powerful interest group in a setting where other legislative issues have an even greater direct bearing on the prosperity of the remainder of economic segments. In the same fashion, senators tend to concentrate on areas of greatest direct benefit to their states despite the fact that all legislation is in some way relevant to all states.

Admittedly, this position must be treated as an alternative hypothesis since the results reported earlier may be reinterpreted so as to be consistent with either the pluralist or elitist schools. For example, the latter may claim that large businesses will support military expenditures, even when not directly beneficial, in order to provide a military umbrella for overseas investment. The hypothesis developed here of a "compensating" strategy and its application must be viewed as an initial effort to take into account some of the difficulties that are faced by an unyielding advocacy of either the pluralist or elitist perspectives.

Unless one is prepared to evaluate existing theories in terms of the results obtained from studies of the empirical reality, a basic issue such as the structure of power in the nation will forever remain beyond the purview of sociology. Rather, we shall continue to be sociologists who subject the topic only to polemics, speculation, anecdotal arguments, and the use of models that would be rejected as "stereotypes" if analogous reasoning were applied to other facets of the social world in which we live.

NOTES

Reprinted from The American Journal of Sociology, Vol. 76, No. 4, (January 1971), pp. 562-584, by permission of The University of Chicago Press.

Ames, William E. 1969. "Seattle Payrolls Fattened by War." Christian Science Monitor, May 23, p. 3.

Baran, Paul A. 1957. The Political Economy of Growth. New York: Monthly Review Press.

Baran, Paul A., and Paul M. Sweezy. 1966. Monopoly Capital: An Essay on the American Economic and Social Order. New York: Monthly Review Press.

Barber, Richard J. 1968. "The New Partnership: Big Government and Big Business," in The Triple Revolution: Social Problems in Depth, edited by Robert Perrucci and Marc Pilisuk. Boston: Little, Brown.

Bell, Daniel. 1958. "The Power Elite--Reconsidered." American Journal of Sociology 64 (November): 238-50.

Benoit, Emile, and Kenneth E. Boulding, eds. 1963. Disarmament and the Economy. New York: Harper & Row.

Blalock, Hubert M., Jr. 1961. "Evaluating the Relative Importance of Variables." American Sociological Review 26 (December): 866-74.

———. 1964. "Controlling for Background Factors: Spuriousness Versus Developmental Sequences." Sociological Inquiry 34 (Winter): 28-40.

Coleman, James S. 1964. "Collective Decisions." *Sociological Inquiry* 34 (Spring): 166-81.

Congressional Index. 1969-70. 91st Cong. New York: Commerce Clearing House.

Fendrich, James M. 1967. "Perceived Reference Group Support: Racial Attitudes and Overt Behavior." *American Sociological Review* 32 (December): 960-70.

Fortune. 1965a "The Fortune Directory of the 500 Largest U.S. Industrial Corporations." (July), pp. 149-68.

_____. 1965b. "The Fortune Directory: Part 2." (August), pp. 169-80.

_____. 1966. "The Fortune Directory of the 500 Largest U.S. Industrial Corporations." (July), pp. 230-64.

_____. 1968. "The Fortune Directory of the 500 Largest U.S. Industrial Corporations." (June), pp. 186-220.

Gracey, Harry L., and C. Arnold Anderson. 1958. "Review of the Power Elite." *Kentucky Law Journal* 46 (Winter): 301-17.

Halverson, Guy. 1969. "Air Power Propels Fort Worth's Upward Economic Spiral." *Christian Science Monitor*, March 28, p. 14.

Horowitz, Irving Louis, ed. 1963. *Power, Politics, and People: The Collected Essays of C. Wright Mills*. New York: Oxford University Press.

Janowitz, Morris. 1960. *The Professional Soldier: A Social and Political Portrait*. Glencoe, Ill.: Free Press.

Joint Economic Committee. 1968. *Economics of Military Procurement, Part 2*. Washington, D.C.: Government Printing Office.

Kokat, Robert G. 1967. "Some Implications of the Economic Impact of Disarmament on the Structure of American Industry." In Joint Economic Committee, *Economic Effect of Vietnam Spending*. Washington, D.C.: Government Printing Office.

Kornhauser, William. 1968. "Power Elite or Veto Groups." In *Reader in Political Sociology*, edited by Frank Lindenfeld. New York: Funk & Wagnalls.

Lapp, Ralph E. 1968. *The Weapons Culture*. New York: Norton.

Leontief, Wassily, Alison Morgan, Karen Polenske, David Simpson, and Edward Tower. 1966. "The Economic Impact-- Industrial and Regional--of an Arms Cut." In *Input-Output Economics*, edited by Wassily Leontief. New York: Oxford University Press.

Miller, Norman C. 1970. "How Educators Build Support in Congress for Fatter School Aid." *Wall Street Journal*, January 20, p. 1.

Mills, C. Wright. 1959. *The Power Elite*. New York: Oxford University Press.

National Resources Committee. 1939. *The Structure of the American Economy. I. Basic Characteristics*. Washington, D.C.: Government Printing Office.

Pilisuk, Marc, and Thomas Hayden. 1968. "Is there a Military-Industrial Complex Which Prevents Peace?: Consensus and Countervailing Power in Pluralistic Systems." In *The Triple Revolution: Social Problems in Depth*, edited by Robert Perrucci and Marc Pilisuk. Boston: Little, Brown.

Rose, Arnold M. 1967. *The Power Structure: Political Process in American Society*. New York: Oxford University Press.

Russett, Bruce M. 1969a. "Who Pays for Defense?" *American Political Science Review* 63 (June): 412-26.

_____. 1969b. "The Price of War." *Trans-Action* 6 (October): 28-35.

Standard and Poor's Corporation. 1965. *Security Owner's Stock Guide, October 1965*. New York: Standard & Poor's Corp.

U.S. Bureau of the Census. 1960. *Historical Statistics of the United States, Colonial Times to 1957*. Washington, D.C.: Government Printing Office.

_____. 1965a. *Historical Statistics of the United States, Colonial Times to 1957; Continuation to 1962 and Revisions*. Washington, D.C.: Government Printing Office.

_____. 1965b. *Statistical Abstract of the United States, 1965*. Washington, D.C.: Government Printing Office.

_____. 1968. *Statistical Abstract of the United States, 1968*. Washington, D.C.: Government Printing Office.

U.S., Congress, Senate. 1969. *Congressional Record*, 115 (March 24): S3072-81.

Weidenbaum, Murray L. 1963. "Problems of Adjustment for Defense Industries." In <u>Disarmament and the Economy</u>, edited by Emile Benoit and Kenneth E. Boulding. New York: Harper & Row.

_____. 1965. "Defense--Space Business." In <u>Politics, Economics, and the General Welfare</u>, edited by Michael D. Reagan. Chicago: Scott, Foresman.

RETIRED MILITARY WITHIN AND WITHOUT
THE MILITARY-INDUSTRIAL COMPLEX

Albert D. Biderman

COSTS

Suppose the Congress next week were to declare Perpetual Peace and the immediate disbanding of the armed forces. It would still find itself obligated a year hence to appropriate a few billion dollars to meet a military payroll for over a million soldiers, sailors, airmen and marines. The obligation would be the contractual one to the retired members of the armed forces. For the remainder of this century and well into the next, there will be a large population of retired professional military men in American society no matter what developments take place in the national or international situation.

A more realistic view than instant disarmament is that we will continue for the indefinite future to make substantial yearly increments to this legacy from the past three decades of mobilization and semimobilization. Without endorsing Defense Department assumptions regarding how much military manpower the nation will decide that it needs in future years, I will accept in this paper, for want of better figures, the official projections of the additional numbers of retirees that our sustained large active duty forces will be generating through and beyond the Year 2000. There are about 800,000 persons now receiving military retired pay. In 1980, the total is projected to reach 1.2 million persons.[1]

We have long been aware that bills for wars come due to be paid by generations unborn at the time the war was fought. There has yet to be such awareness of what the future bills are like for an armed era. I am concerned here with that large part of this bill that involves payments to warriors when they are no longer needed or suited for warfare.

To keep the rank-seniority structure of the military in its traditional tidy pyramidal form, to maintain its youthful vigor, and to keep it up-to-date with regard to its continually changing needs for skills, the large majority of career personnel must be thrust out of the system at a fairly early age. Our armed forces retire the large majority of their career members at, or shortly after, they have served for 20 years, and practically everyone before they have exceeded 30 years of service. There is recognition that a person who has devoted the major portion of his adulthood to learning the arts of war may be substantially handicapped in assuming other remunerative arts when he leaves the armed forces. Since the average career man is in his mid-forties upon retirement, with almost half of his working years still before him, however, the system assumes that he can meet the major part of his money needs by second-career work. Only relatively modest subsidization by retired pay should be needed to remedy the competitive handicap he suffers by virtue of his very late entrance into the civilian job market.

A full pension system would, in any event, be prohibitively costly. As it is, the costs of retired pay have become so staggeringly high that both the Congress and the Executive are beginning to feel some urgency about finding some alternative to it. The costs of retired pay crossed the $2 billion mark in 1968, reached $3.9 billion in 1971 and were projected to reach $4.8 billion in 1973.[2]

In terms of a purely political calculus, contrasting the retired-pay appropriation with other items of the federal budget highlights its political unattractiveness. Retired pay in FY 1968 was about one-tenth the amount needed to pay every one in uniform. It would have been even higher, relatively, had not the Vietnam War swelled the active duty ranks.

Furthermore, the cost of retired pay is increasing disproportionately, so that by 1980, expenditures for retired pay are expected to be one-fifth the total budget for military personnel, and by the year 2020, almost 30 per cent. And this, of course, involves Defense Department projections of active duty manpower needs, not those of arms control advocates.

In pay alone, the retired military required greater expenditures in 1967 than the total expenditures of each of 4 of the 12 federal departments. Military retired pay was some $140 million greater than that for OEO's poverty program. It cost four-and-a-half times as many federal dollars as those required for all operations of the State or Labor departments.[3]

Looked at with a political eye cast toward constituencies, the retired military are just a handful of men as compared with the over 25 million veterans in the population. But in 1970 retired pay was an amount equal to almost one-third of the entire expenditure of the Veterans' Administration.[4]

The costs of retired pay constitute a special thorn in the side for those who worry about the federal budget because the amount needed for retired pay must be appropriated anew each year. Unlike funded plans, the annual appropriation character of retired pay gives it special and painful visibility. Had a fund continually been put aside which, actually or theoretically, accrued interest during the entire period of the active duty of these men from which payments could be made upon their retirement, the bite on the public purse would not appear as severe as it does at present. Provision will almost certainly be made in the near future for a funded, contributory system. Such a system will provide bookkeeping memoranda that will make clear to military retirees the current income they forego in the interests of future financial security and that will remind the country of the future money obligations it is accruing by keeping men in uniform now. Current official proposals would also bring financial provisions for former military personnel into more realistic alignment with the facts of second career patterns.[5]

From the standpoint of future costs, current proposals look forward to sharply reducing the proportions of military personnel that serve until retirement eligibility. Vested pension rights would contribute to this end. Although reductions in the absolute size of the armed forces are the surest route to future savings, the current plans for all-volunteer force can complicate the problem by increasing reliance on long-term careerists, rather than temporary soldiers. Regardless of how much we curtail the flow of retirees into the civil society in the future, however, only the very slow processes of aging and mortality will erase the commitments to pay retirees that exist already.

JOBS

While those with an anti-military animus find these large bills for ex-military professionals disconcerting, they are more vocally alarmed by the one thing that keeps these costs from being yet higher -- that is, the remunerative employments of the retirees. The majority of retired military meet most of their financial needs by earnings from civilian jobs. Labor force participation rates for recent yearly retirement cohorts exceed 90 per cent.

As of 1966, when the most recent survey of military retired employment was made, the second-career system was working fairly well. Although earnings of retirees averaged considerably below those of civilians of comparable age and educational attainments, and much below active duty income, retired pay usually more than made up the difference. Depending upon their ranks and years of service, retired officers' total income from all sources after retirement averaged from 9 to 19 per cent more than their active duty income, and enlisted men, on the average, bettered their active duty income by from 6 to 40 per cent, according to Defense Department calculations. But these figures were based only on officers who had completed four years of college and on enlisted men who had completed high school.[6] Among the substantial minorities who experienced income <u>losses</u>

after retirement, those who hadn't crossed these crucial educational hurdles bulked largest.

The favorable averages mask minority hardship. Unemployment and underemployment rates among military retirees in the mid-1960's were consistently more than twice those of comparable civilian age groups -- and if the comparisons were restricted to white males -- the difference is even more extreme.[7] To be sure, the cushion of retired pay makes the penalty of unemployment less severe for the military retiree than for the civilian.

Retired pay is a blunderbuss device for overcoming the sacrifices of earnings potential involved in military service. Such sacrifices obviously vary widely with the specific character of the career pursued in service, while retired pay rates vary only by rank and length of service. Often, service experience may be an occupational advantage for which one should have been willing to pay a great deal. A system of variable pension depending upon the civilian-job relevance of service experience is conceivable, although the data-requirements for imparting equity to any such system are formidable. As it is, variability in potential income sacrifice makes it inevitable that any blunderbuss income-maintenance system must make average payments which substantially exceed average sacrifice if it is to succeed in the objective of avoiding extreme hardship among substantial minorities. The tolerable size of such minorities is critical for evaluating the success of the retirement system.

IMPACTS

Although the imminent prospect of over 1,000,000 former military professionals in the civilian work force may appear formidable, the social impact of this development will probably be a relatively modest one.

Retired professionals are a small group relative to the great number of Americans who have had military service -- they constitute only about 3 per cent of all war veterans. But such impact as they do have on the civil society is presumably more distinctively a product of their military lives than in the case of the ordinary veteran.

Of course, 1,000,000 men is absolutely not a small number of persons. To suggest the order of magnitude involved, there are already about as many retired military personnel in the labor force as there are engineers of all types, there are several times as many as there are butchers, bakers and candlestick-makers, combined.

HOMOGENEITY

We know, however, that the military calling is occupationally far more heterogeneous than butchers, bakers, or candlestick-makers, or even than engineers; indeed, it encompasses all these occupations and many more. The callings of the retired in their civilian lives become yet more varied. Nor can we assume that the interests and ideologies representative among the military population are as homogeneous as those of any given occupational group -- that members of Retired Officers Association are as like-minded or more so than, say, members of the United Steel Workers or of the American Psychological Association. An important question that remains to be answered in assessing its impact is the actual degree of homogeneity and concert of this population element.

INFLUENCE

The high-ranking officers who achieve positions of great influence and prominence in civil life are a very tiny fraction of the retiree population. Enlisted men form a large and growing majority of all retired personnel and retired officers are predominantly in the middle (major -- lieutenant colonel) ranks. To the extent that the general status levels of military and second-career jobs can be compared, the number who move downward or laterally on retirement far out-balance those moving upward. The civilian organizations into which the retiree usually moves, furthermore, operate on a far more modest economic scale than the military organization from which he comes. In the positions they come to occupy when they leave the service, the retirees, high ranking or low, usually occupy positions in which they each have at their command far less of the material and human capital of the national economy than they ordinarily had in their military jobs.

On a national scale, retirees are not now nor are they destined to be numerous enough to be an important economic factor in the national economy. Government is the only broad employment sector in which retirees constitute a major source of economic competition with civilian workers. Government is also the class of employer most often in a position to make direct use of special skills and experience commonly found among former military professionals. But in 1966, the most recent date for which the computation can be made, retirees comprised only about 3 per cent of the federal payroll.[8]

In private business as well as in government, a large, although not accurately assessable proportion of the jobs filled by military retirees serve the defense complex. This is not only because large numbers of retirees have Defense Department civilian jobs and jobs in defense industry, but also many of their ostensibly nondefense jobs service this sector indirectly. Retirees settle predominantly in areas in which defense is a major industry, if not the major one, and particularly in areas having large military installations.

To the extent that military retirees are concentrated in localities and in organizations already in the defense orbit, however, they have added proportionately little additional militarizing influence to the military influence already present in these communities. In cities where they are most numerous, such as San Antonio, San Diego, and various Florida communities, they are outnumbered several-fold by the active-duty military in the area. The Washington metropolitan area abounds with retired brass, but the brass in the Pentagon has a more powerful glare. What is true, however, is that there will be communities in the future with substantial military-oriented populations, no matter how greatly active duty components are reduced. To the degree that retirees continue to concentrate in a relatively small number of communities, one would expect their influence, as well as their visibility, to be greater than would be the case with dispersion. There are reputed to be 500 retired generals in the San Antonio area, for example. There are over 25,000 military retirees in the National Capital area. Washington-area retirees are drawn particularly from those who have had repeated assignments in headquarters positions. In addition to their being politically experienced, atypically high ranks and location at the political center make this group of retirees at least potentially of political importance.

Although several individual ex-military professionals have had political prominence in recent years, retirees as a group have not constituted a significant political force. Political activity by retiree organizations has been limited largely to matters of direct material concern, such as retired pay legislation. There appears to be important carryover of the doctrine of apolitical neutrality from the active duty to the retired status. As a consequence, the high ranking officer who gains prominence in the national political scene usually does so in a rather politically gray, elder-statesman role, epitomized in recent years by Gen. Lucius Clay. Deviance from the apolitical stance by entrance into electoral or mass-movement politics usually is associated with political deviance, as well. The deviant ideological politics of General Edwin A. Walker or Captain John G. Crommelin or General Curtis LeMay are illustrative.

Janowitz[9] has observed that the number of retired military in key executive and administrative posts in the nation was at a peak in the period following World War II. Our subsequent wars have not served as well to establish prominent public reputations for military leaders. Barring some spectacular feat of historical revisionism, this should prove particularly true of the Vietnam war.

The transient lives of military men, and their national, rather than localistic orientations, are not conducive to the formation of strong positions in the state and local political apparatuses that, in our federal system, are the usual entry points for the pursuit of political careers. This may account for the fact that it has been almost as likely at any time to find a retired officer in the White House as in the Congress. Only two regular retirees have served in Congress in recent years. Col. Frank Kowalski, Jr. served as Congressman-at-Large from Connecticut in the 86th and 97th Congresses and Captain William R. Anderson currently represents a Tennessee district. Notably, both gained reputations as mavericks and both gained prominence for attacks on aspects of the military status quo.

Nonetheless, there is doubtless considerable homogeneity of political outlook among the majority of retirees. The mobilization of retirees for political activity of a rightist character is conceivable under conditions where personally experienced disadvantage was seen by them as of a piece with fundamental political decadence in national affairs. There are incipient themes of this character in the media directed to retirees.

CONFLICTS

Those who would restrict the impact of the military on the

society confront dilemmas in the case of the military retiree. On the one hand, there is concern because so large a number are moving from the military into the civilian sphere, bringing with them unwelcome influences from the former on the latter. On the other hand, there are concerns which arise because many retirees remain within the defense complex in their second career. They thereby constitute an added source of difficulty for attempts to reduce the size and influence of the "military-industrial complex." Defense employment of retirees also generates alarm concerning conflicts of interest -- the prospect that the military man will cater to private interests while on active duty in order to gain preferment in employment upon his retirement and of his improperly using information and personal contacts in postretirement defense industry jobs.

Gary Spencer, who investigated the second careers of retired West Pointers, is illustrative.[10] He found that they were employed most frequently by large Defense contractors (31%), by educational institutions (also 31%) or by financial organizations (15%). He had equal dismay to express about the jobs outside the defense complex taken by those who had "... spent the past twenty or thirty years in caste system being saluted and obeyed without question..." as about the jobs taken within the defense complex.

Although the second-career patterns of West Point graduates are quite untypical with regard to high numbers taking jobs in schools and colleges and with the major defense contractors, my own studies suggest that there has been as great and perhaps an even greater proportion of all retirees, enlisted as well as officer, that has been dependent on defense-generated employment for second-career jobs. So much so that I have some doubts about the viability of the second-career system without a military-industrial complex to support it. This is not because I find the facts in accordance with the impressions created by Senator Proxmire's recent investigations -- that is, that the largest defense contractors were the major employers of ex-military. My calculations from Senator Proxmire's figures indicate that about 8 per cent of higher ranking regular retirees (colonel or general officer) were employed by the 100 largest defense contractors and, if lesser contractors employed such men in the same proportion to their dollar volume, total employment in the defense contractor universe would be about 12 per cent. Employment in defense industries of retired reservists and lesser ranking regulars seems somewhat less frequent (I can make no quantitative estimate).

While the statement "2,000 Ex-Brass Serve Defense Contractors" serves excellently for making "Ah, hah!"-type headlines, it does so because of what it implies rather than what it reveals. Judging from data on salaries of retired senior officers employed in defense industry, few of them hold jobs that are sufficiently well-paid to suggest that peddling highly valuable inside dope or influence is involved.[11] Just how misleading the criterion of numbers can be was shown by testimony regarding retiree employment in the firm which stood out in 1958 as the greatest hirer of retired officers in relation to its volume of contracts. The highest paid officer in the firm, testimony revealed, was a retired admiral at $14,000/year and the median pay of all retired officers the firm employed was $650/month. A third of the officers held blue-collar jobs.[12]

The 100 largest defense contractors include firms that account for much of the nondefense economy of the nation, as well. Any retiree employed by ATT, General Motors, ITT or any one of several conglomerates is eligible for the list, whether his work has anything directly to do with defense or not. To date, there has been no systematic evidence developed on the extent and nature of activities of retired officers as a cement binding together the military and industrial segments of "the complex." The irrelevance of the employments of thousands of retired officers for the issues posed by the concept of a "military-industrial complex," however, does not dispose of the matter.

The roles of a few score key figures are indeed more pertinent than are those of the "rank and file" retired colonel or general (there were in 1970 about 40,000 retirees at the 0-6 level and 3,700 in the general and flag ranks).

Attempts by Congress to legislate against improper business activities of retirees have focused on "selling" to the armed forces. The nature of defense production and procurement is such that the concept of the man with an order book, or even that of the bid-writer or proposal-hustler, can cover only a minor portion of all the activities in which prior knowledge and current contacts of military men can lend sales advantage to a firm. It is hardly surprising that in the entire history of the laws against selling, they have rarely been invoked and, in the two cases in which adverse judgments against retired officers were made, the cases were subsequently overturned.[13] The doctrine of conflict of interest becomes logically extended, as Rep. F. Edward Hebert would have it, to mean that it is "...unethical and unconscionable for a person to have anything to do in private life with a subject with which he was directly concerned while in public employment."[14] The purpose of any activity in a commercial organization is to contribute to sales. And, in the "contract state," there are few large organizations which are not in one way or another involved in doing business with the government. Complete control by statute of "conflict of interest" would have to ban not only "selling," but all employment with firms doing business with the armed forces.

Although the implications of ill-deserved preferment and private gain helps lend emotional tone to publicity about defense employment of ex-military, it is a subordinate issue for those who worry about the "military-industrial complex." As Sen. Proxmire put it:

> This danger does not come from corruption. Except in rare circumstances this is no more prevalent among military officers than among those with comparable civilian responsibilities.
>
> The danger to the public interest is that these firms and the former officers they employ have a community of interest with the military itself. They hold a narrow view of public priorities based on self-interest. They have a largely uncritical view of defense spending....
>
> In too many cases they may see only military answers to exceedingly complex diplomatic and political problems. A military response or the ability to make one, may seem to them the most appropriate answer to every international threat.[15]

As compared with the situation in defense industry, there are stronger grounds for the complaint of the civil servants unions about unfairly subsidized and advantaged competition for their jobs from military retirees. In 1966, one-third of all retired enlisted men and reserve officers had held federal civil service jobs since their retirement, as had about 20 per cent of retired regulars. Subsequently, civil service employment has become even more frequent among regular officers. Of federally employed retirees, almost two-thirds were employed in defense agencies.

Employment in industrial or government defense work does not exhaust the dependence of retiree second-career employment on defense. As noted earlier, retirees settle predominantly in areas in which defense is a major industry, if not the major one, and particularly, in areas having large military installations. Many of their ostensibly nondefense jobs in these areas are service jobs which have as their sole or major clientele defense organizations and their personnel.

All in all, I would estimate that from 50-to-60 per cent of retiree employment has been quite directly dependent upon the defense and aerospace economy.

Both formal rules and informal discrimination restrict retiree employment now as in the past, notably in the civil service or in selling to the services. Were there to be significantly greater impediments to the hiring of retired military, however, it appears likely that large numbers of military retirees would not be able to hold their own economically. Indeed, it is probably the case that there will have to be greater facilitation of second-career opportunities in the future, if financial hardship is not to become dangerously common among the retired military. The most important substantial statutory changes in recent years have, in fact, operated in the liberalizing direction. These were enacted in 1964. They allowed retired regular officers to hold civil service jobs and changed the formulae which set trade-offs between retired pay and earnings from federal "second career" employment in a way which bettered the terms for the higher paid retirees. This was immediately reflected in larger percentages of higher ranking officers taking civil service jobs.

The success of the second-career system as of the time it was last assessed in 1966 appears to have been made possible by a full-employment economy, and, particularly, an economy with substantial and thriving defense and aerospace sectors. We have no direct data for evaluating what hazards may exist now that neither of these conditions exists, or what they might be in the future, as we continue to have major influxes of retirees into a difficult employment market with a declining defense sector.

Retirees have been able to achieve economically and subjectively rewarding second careers, particularly in middle and lower managerial jobs in large organizations serving public purposes. The national commitment to defense economy has created such jobs, has validated the legitimacy of employing former military personnel in them, and has given retirees the sense of contributing to valued social purposes. Sharply reduced national commitments to defense would not only immediately restrict jobs available for retirees but also could affect their acceptance by employers and their esteem.

PUBLIC SERVICE

Probably more than in the case of other workers, the situation of military retirees in the event of radical curtailment of defense expenditures will depend upon the growth of other public programs. Retirees have already constituted a useful manpower resource in a number of areas of public service where problems are becoming increasingly critical. These include technical training programs, medical services, and police and other protective services.

CIVIL SERVICE

With over 80,000 military retirees in civil service positions and with apparently increasing proportions of retirees going into federal service, the impact of retirees on civil government may appear impressive. Even here, however, their numbers are actually modest when viewed in relation to the size of the government establishment -- as noted earlier about 3 per cent of the total federal payroll. About two-thirds of the retirees are employed in defense agencies, as compared with 41 per cent of all civil servants. Even in defense, however, their numbers are small relative to the total civilian component -- about 6.5 per cent. In no other federal agency are retirees as high a percentage of the employees.[16]

These simple comparisons, it is true, somewhat understate the significance of military retirees in the federal government in that they are represented in somewhat greater proportions in the middle and upper grade levels of the civil service. Retirees in the federal service also tend to have more important positions than those in the private sector. For example, in 1966, about 31 per cent of the retired officers working for the federal government held professional or technical jobs, as compared with only 19 per cent of all employed retired officers. But again, the professional level appointments were mostly in defense agencies.[17]

The major significance of the civil service for the second career problems is that of a large and expanding employment market for the lower level, and even more particularly, for middle-ranking retired military personnel. In addition to the special features which make government jobs more congenial to people with military backgrounds, the civil service is relatively free from such barriers to entry as rigid seniority systems, closed unions, and nepotistic hiring.

As a consequence, the second largest government employer of retired military is the Post Office department. As a vast hirer of lower-level administrative, clerical and blue-collar personnel, it affords jobs for 12 per cent of the federally employed officers and 21 per cent of those who retired as enlisted men. If we add together all low-level categories -- retirees working in the postal field service, those in the lower-level General Service "white collar" categories, GS 1-7, and those holding blue collar "Wage Board" positions -- the bulk of all retiree civil service employment (70%) is accounted for. Even considering retired officers, alone, almost 40 per cent are in these categories.[18]

Janowitz[19] has proposed formalization of the circumstances which already make public service employment a natural and most frequent second-career route for military retirees. Those who serve short periods in the prospective all-volunteer force, as well as those who continue in the military for the periods which currently make for retirement eligibility, could consider themselves as embarking on a "lifetime career in the public service ... with the military being the first portion and a second portion in the civil establishment." He contemplates arrangements covering not only federal service but state and local employment positions as well, through negotiated agreements between the national and state and local governments. Presumably, such a system would also serve for flexible lateral movements between uniformed and civilian roles, as national requirements changed. Such a plan, Janowitz reports, is already operating with effectiveness in the Federal Republic of Germany.

A number of elements of this kind of system would not be novel in the United States, although existing encouragements of military-to-civil service careers stem more from the favor accorded by civil service systems to the broad class of war veterans than to the special class of career military personnel. They include such things as veterans preference in hiring and some counting of years of military service for eligibility for civil service perquisites. Full and formalized adoption of the Janowitz type of plan, however, would involve, for better or worse, radical blurring of lines between the civil and military components of government.

State and local government, which have had a dramatically growing labor force, present what appear, on their face, to be many particularly good matches to common skills and aspirations of military retirees. As of the most recent surveys, however, few retirees were employed at the lower levels of government. Presumably, seniority rules afford some impediments to their hiring, but the relatively tenuous local ties and information of most retirees probably are greater sources of difficulty. Employment in higher and secondary education and in protective services are, to some extent, exceptions in employing fairly substantial numbers of retirees. Even in these two fields, however, employment rates are lower than would be expected given the experience, skills and preferences of retirees.

Police work would appear to be one of the most suitable types of job, but the general lack of provision for lateral entry into large police systems is a major barrier. If this condition persists, the switch of the talents of military men from international to domestic warfare will continue to take place in large measure within the expanding private protective services industry.

CONCERN WITH RETIREES' EMPLOYMENT PROBLEMS

Mass media attention to retirees, including the most publicized expressions of the concerns of Congress, has been restricted almost exclusively to alarm about illicit and dangerously influential employment of former officers. While public attention is sporadically drawn to these concerns about the jobs retired military are getting, official attention is paid increasingly to concern about jobs retirees fail to get. Unemployment, underemployment, low-income employment and ill-suited employment affect large proportions of the retirees. There are many indications that, with the increase in the retiree population, conditions are becoming worse rather than better. The dangers to the retirees are not only that they may be competing increasingly with each other for the same kinds of jobs in the same kinds of localities in ever-tightening special labor markets, but also that political pressures against their employment will mount. The unpopularity of the military during an unpopular war doubtless facilitates such reactions. Defense industry, some spokesmen of which have already indicated their wariness toward hiring retirees because of public relations problems, may have become increasingly wary after the campaign conducted by Senator Proxmire in 1969. Publicized opposition of organized civil servants to retiree hiring by government also became more vehement at about the same time.

The United States has yet to experience a demobilization of its career forces on the same order of magnitude as has been the case in some other countries. Britain, France and the Low Countries experienced an identical problem to that of the United States in having a World War II-generated hump of retirement eligibles in the late 1950's, but, in addition, they undertook major reductions of their military forces when they dissolved their empires. Each country had to institute special programs for accommodating retirees in the civilian labor force. An attempt by the Soviet Union in 1960 to handle its hump problems and forces reductions by the retirement of large numbers of career personnel to such unattractive employments as Khrushchev's

Virgin Lands program apparently generated such great backfire from the military as to force a reversal of the program. The military-political problem in France was so entangled with other aspects of the liquidation of the empire as to offer only dubious parallels to any prospective developments in the United States. Nonetheless, the objective situation of our own military is such that the fortunes of its members, after as well as before retirement, are being directly affected by our equivalent to the surrender of imperial dominions.

The traditional solution to the problem of civil control of the military in the United States has been to have a very small professional military establishment and to keep military professionals sharply separated from the civil society. Neither condition now exists. New and more elaborate forms for maintaining civil supremacy have had to be found. This is also true with regard to the problems considered here. In the United States, unlike the case in many other countries, there has never been a large military class threatened by, or actually experiencing, severe problems of economic and status loss. Under such circumstances, an aggrieved military class might be impelled toward political involvement. For this reason, even those who most fear military influences on the civil society may see a stake in the occupational accommodation of retirees. Even with miraculous peace and instant disarmament, the question can no longer be one of excluding military men from major participation in civil institutions, but only one of the nature of the participation they are to have.[20]

NOTES

[1] Information on the size and cost of the retired military population is based on data provided by Mr. Joseph Glenn, Actuarial Consultant, Office of the Assistant Secretary of Defense (Manpower and Reserve Affairs).

[2] Special Analyses. Budget of the United States Government, (Washington, D.C.: U.S. Government Printing Office, 1972), p. 200.

[3] Military retired data from J. Glenn, OASD (M&RA). Comparative data on budget outlays of nonmilitary Federal Departments and agencies are from U.S. Bureau of the Census, Statistical Abstract of the United States 1969 (Washington, D.C.: U.S. Government Printing Office, 1969), pp. 330 and 379.

[4] Military retired data from J. Glenn, OASD (M&RA). Data on expenditures for veteran's benefits from Statistical Abstract of the United States 1971, pp. 261-262.

[5] The Defense Department's "Hubbell Committee" (so-called after its Director, Rear Admiral L. E. Hubbell, U.S. Navy), which conducted the Congressionally-mandated First Quadrennial Review of Military Compensation, for example, recommended a funded, contributory, two-step annuity plan. The plan, predicted on the average, second-career income loss of retirees surveyed in 1966, would provide a lower level of retirement pay until the end of a presumed normal working life and a higher level of pay in old age. U.S. Department of Defense, Modernizing Military Pay, Vols. IV and V, January 15, 1969.

[6] Ibid., Vol. IV., p. 2-9.

[7] Based on military retiree unemployment rates from unpublished Hubbell Survey data and civilian unemployment rates from U.S. Department of Labor, Handbook of Labor Statistics 1969 (Washington, D.C.: U.S. Government Printing Office, 1969), p. 107.

[8]Based on unpublished Hubbell Survey data. See also, U.S. Civil Service Commission, "Report on the Survey of Members of the Uniformed Services, June-July, 1966 for the United States Senate Committee on Post Office and Civil Service," November, 1966.

[9]Morris Janowitz, The Professional Soldier: A Social and Political Portrait (New York: Free Press, 1960), p. xxvii.

[10]Gary Spencer, "Second Careers of West Point Officer Retirees," paper read at the annual meeting of the American Sociological Association, Denver, Colorado, September, 1971.

[11]Albert D. Biderman, "The Retired Military," in Roger W. Little (ed.), A Survey of Military Institutions, Vol. II (Arlington, Va.: U.S. Air Force Office of Scientific Research, 1969), pp. 564-565. Also in Roger W. Little (ed.), Handbook of Military Institutions (Beverly Hills, California: Sage Publications, 1971), p. 157.

[12]U.S. Congress, House, Hearings before the Subcommittee for Special Investigations of the Committee on Armed Services, Retired Military and Civilian Personnel by Defense Industries, 86th Congress, 1st Session (Washington, D.C.: U.S. Government Printing Office, 1959), pp. 535-543.

[13]Biderman, "The Retired Military," op. cit., pp. 563 (Survey), p. 155 (Handbook).

[14]U.S. Congress, House, Subcommittee for Special Investigations of the Committee on Armed Services, Employment of Retired Commissioned Officers by Defense Department Contractors, 86th Congress, 1st Session (Washington, D.C.: Government Printing Office, 1960), p. 20.

[15]U.S. Congress, Senate, Congressional Record, March 24, 1969, p. S3074.

[16]Based on unpublished Hubbell Survey data.

[17]Ibid.

[18]Ibid.

[19]Morris Janowitz, "Adapting the Armed Forces to an All-Volunteer System: Recommendations for a Democratic Society," unpublished manuscript, n.d., pp. 5-7.

[20]In addition to the sources already cited, this paper draws on previous research undertaken or participated in by the author. Important insights into the postretirement experience of retired military professionals emerged from a sample survey conducted by

Laure M. Sharp and the author in 1964. Much of the author's research in this area was conducted in connection with the "Military Establishment Project" of the Twentieth Century Fund or was supported by the Inter-University Seminar on Armed Forces and Society. Fuller discussions of some of the topics discussed here may be found in:

Biderman, Albert D. "The Prospective Impact of Large Scale Military Retirement." Social Problems (Summer 1959) 7:84-90.

_____. "Sequels to a Military Career: The Retired Military Professional." In Morris Janowitz (ed.), The New Military: Changing Patterns of Organization. New York: Russell Sage Foundation, 1964.

Sharp, Laure M., and Biderman, Albert D. "Out of Uniform: The Employment Experience of Retired Servicemen Who Seek a Second Career." Monthly Labor Review (January 1967) 90, 1:15-21.

_____, and Biderman, Albert D. "Out of Uniform: Educational Attainment Seen as a Key Factor for Retired Servicemen in the Establishment of a Second Career." Monthly Labor Review (February 1967) 90, 2:39-47.

Biderman, Albert D., and Sharp, Laure M. "The Convergence of Military and Civilian Occupational Structures: Evidence from Studies of Military Retired Employment." American Journal of Sociology (January 1968) 73:381-399.

Biderman, Albert D. "Relationships Between Active Duty and Post-Retirement Careers." In N. A. B. Wilson (ed.), Manpower Research. London: English Universities Press, 1969, pp. 426-441.

Bruce B. Dunning assisted in the preparation of the present paper.

Part II

STRATEGIC ALTERNATIVES

STRATEGIC DIMENSIONS OF AN ALL VOLUNTEER ARMED FORCE

Morris Janowitz

The mass conscript armed force with its vast mobilization reserves is currently being phased out of existence in the NATO nations. This event is certain to have a profound effect upon international relations, as well as internal relations between the civil and military sectors within the NATO countries. International relations specialists are understandably reluctant to confront the decline of mass armies in the affluent nations of the West, a decline which is directly linked to advanced industrialization (though not caused by it). As late as December 1970, in the prestigious Adelphi papers of the Institute of Strategic Studies, Erwin Hackel, a young political scientist from the University of Konstanz, concluded that "there is no easily discernible trend in the present debate" on military manpower systems in NATO countries. Yet the movement away from conscription and toward a greater reliance upon an all-volunteer system or upon a mixed militia system and professional soldiers and short term conscripts was already under way.

In the United States, one campaign appeal that President Nixon sought vigorously to implement after taking office in 1969 was to end the draft as quickly as possible and to create an all-volunteer force. He established his President's Commission on an All-Volunteer Armed Force not with the mandate to explore alternative formats but to make specific recommendations and estimates of costs for ending Selective Service before the next presidential election. Paradoxically, the prolongation of hostilities in Vietnam only served to speed up the end of conscription and develop congressional support for his campaign promise. The termination of conscription was one issue upon which anti-war congressmen and pressure groups could unite with the Nixon administration. The result was the effective political decision not to extend Selective Service legislation beyond 1 July 1973, and the initiation of planning by military officials to reach the objective of a "zero draft call" by 1 January 1973 so that there would be a six-month period of accommodation and transition.

The ending of the draft in the United States will have a deep impact on military manpower systems in Western Europe. Rather than working to maintain existing conscriptions systems, it will tend to push NATO nations toward an all-volunteer system or toward new forms of militia systems. Over a decade ago, in 1960, Great Britain introduced an all-volunteer system, and the 1970s will certainly see further overall reductions in British military manpower because of economic pressure and the sheer difficulties of recruitment. In the last three years NATO countries have either reduced the length of conscript service or are debating such reduction. More radical measures are certain to be examined closely after the end of the draft in the United States. The Netherlands, with its powerful commitment to NATO principles and strategy, is openly debating the conditions under which it will institute an all-volunteer armed force and actively planning such a system. Norway, with its reserved NATO status, may well continue the draft longer as a demonstration of its emerging detachment. In Germany, Helmut Schmidt, the Socialist minister of defense, has advocated an all-volunteer cadre augmented by a short-term (six-month) conscript militia. In Italy, and to a lesser extent France, the size and nature of the manpower systems are concrete not only to international relations but also to internal security, and consequently debate on shifting toward a more voluntary force has been retarded.

Can an all-volunteer system produce an armed force of sufficient strength and quality for an effective international posture of deterrence? The prospect of an all-volunteer armed force also causes deep concern about civil-military relations, especially with regard to the question of social isolation or even "alienation" of the military from the larger civilian society. In the United States the military establishment, and the ground forces in particular, are experiencing a profound crisis in legitimacy; this is due in part to the impact of the agonies of Vietnam, but it involves also internal race tension, corruption, and extensive drug abuse, disintegration of command and operational effectiveness, widespread antimilitary sentiment, and a continuous reduction in force levels which limits career opportunities.

Many high-ranking military officers in the United States at first viewed the end of the draft with utter dismay. Their outlook, of course, reflected their concern with the prerogatives and resources of the military, but they generally felt also that the draft was essential to maintain required manpower levels and that it was politically and morally undesirable for a democratic society to rely on an all-volunteer force. However, as internal tensions in the armed forces have become progressively more disruptive, the established command has come to look toward an all-volunteer system as an acceptable outcome. Younger and more innovative officers, in fact, see the advent of the all-volunteer force as an essential precondition to massive internal professional reform, a first step in reconstructing the ground forces, regardless of the strategic limitations it will impose.

DECLINE OF MASS ARMIES

The military establishment, in any historical period, is both a reflection of the larger society and an institution in its own right with a distinctive environment and ethos. Thus, the ending of draft in the United States is a dramatic historical transformation of American society taking place in our own time. It also makes another step in the end of an epoch in the rise and decline of the mass armed force in world history.

The mass armed force had its origins in both technological and sociopolitical change. On the one hand, the mass army was rooted in an organizational system created by increased firepower of the infantry and artillery plus improved means of transportation of personnel and supplies. Historical epochs do not of course start or conclude on specific textbook dates. But the technology of the mass army was certainly in operation during the American Civil War and the Franco-Prussian conflict, and essential prototype elements, especially its organizational features, were already in existence during the Napoleonic wars.

On the other hand, there are strong reasons to trace the origin of the mass armed force to the sociopolitical struggles of the American and French revolutions and the forms of modern nationalism which they produced. These rebellions marked the end of the postfeudal armies as the revolutionary leaders armed the ordinary citizenry. The idea that citizenship involved the right and duty to bear arms -- truly a revolutionary notion -- came into being. In fact, military service was an essential element in establishing and expanding the scope of modern citizenship. To be a citizen of the nation-states was to have the right and duty to bear arms in defense of the state. (It is striking to recall that during World Wars I and II, elements in the black community in the United States demanded the right to serve in combat units as an expression of their aspiration to full citizenship.)

In Europe, after the French Revolution, the mass armed forces developed professional cadres which were augmented by a conscript and mobilization system. Although the institution rested on an ethos of citizen participation, the professional officers were in actuality highly distinct from the rest of society. (This was not the first time in the history of political and institutional change that a protest movement produced unanticipated consequences.) In the United States, the professional cadres were smaller, and the mass armed force did not become effectively institutionalized until the end of the nineteenth century.

In spite of their revolutionary origins, the professional armed forces ended by serving the cause of nationalism very well: the officer corps of Western Europe had no difficulty in transferring its feudal-based allegiance to the modern bureaucratic nation-state. A corresponding process took place in the United States. Mass armies supplied an opportunity for the lower classes to participate directly in the national polity in a manner they could readily manage and appreciate. Service in the conscript forces in the nineteenth and twentieth centuries was for a significant segment of the population -- even after the slaughter of World War I and up through World War II -- an act of political affirmation. In both Europe and the United States it became an expression of popular nationalism, undercutting, in Europe, the countermovements toward internationalism and socialism. The right and duty to participate in the conscript armed force, as much as the extension of the franchise, was at the core of the political transformation of modern nationalism.

The distinctive, professional officer corps of the mass armed force, with its strong sense of separation from civilian society, in due course brought with it its own elements of institutional transformation, including an increase in sheer size. In Europe since the close of the Franco-Prussian War and in the United States since the mobilization of World War I, a dominant trend in the mass armed force has been toward "civilianization" of the military. Preparing for war and making war give the military its distinctive institutional climate. However, the boundary between military forces and civilian society has weakened as total mobilization requires larger and larger segments of the population to become part of the war apparatus. Air warfare has meant that the entire population is a target for military action. Military leaders must share authority with civilian scientists as technology becomes more and more complex, and the influx of civilians into the officer corps during periods of mobilization undermines traditional forms of authority and command. Within the professional military, the source of social recruitment into the officer corps has broadened, the concentration of personnel with civilian-type skills has increased, and the patterns of military authority has shifted from authoritarian command to organizational decision making.

The process of civilianization of the mass military is not simply an outgrowth of technology and organizational control. The vast resources required for military operations and the need to justify prolonged hostilities and massive destruction necessitate egalitarian ideology, both in democratic and totalitarian societies. Increasingly, men are no longer prepared to fight for nationalist sentiments alone; the cause, rather, must be seen as justified morally. Military institutions require direct civilian control for legitimacy. Although the expanded resources of the military permit it to operate as a very powerful pressure group, the threat in advanced industrial nations of old-fashioned military dictatorship seems remote.

The decline of mass armies in the affluent nation-states of the West began after World War II, although it has taken twenty-five years for the process to become fully self-evident. Again, both technological and sociopolitical factors initiate the change. Deployment of nuclear weapons marked the technological transformation of the armed forces of the NATO nations as the sheer destructive power of these instruments altered the scope of war making. But the introduction of nuclear weapons did not per se make inevitable the gradual erosion of mass armies; it was only a precondition.

In advanced industrialized societies the purpose of military institutions in general has been subjected to massive criticism, and in particular the moral value of conscript service has been shaken. Hedonism, personal expression, opposition to the life style of the military establishment, and resistance to military authority, as well as a pervasive moral criticism have become paramount among young people. The destructive potential of nuclear weapons has served not only to produce moral opposition to violence, as well as new forms of pacifism, but to heighten realistic understanding of the effective interdependence of national societies. The use of force has traditionally operated within circumscribed limits; the new moral and political definitions serve to generate a powerful sense of neutralism. Literacy, patterns of mass consumption, and political rhetoric have superseded military service as the hallmark of citizenship. Nationalism itself is muted and mixed with diffuse but powerful feelings of transnationalism. The campaigns of the United States forces in Southeast Asia, of course, supplied in the United States an emotional basis to the emerging popular pacificism. These trends are concentrated among an important minority of young people, but can be found in varying degrees in all parts of the social structure. Thus, in Germany, reluctance to serve in the armed forces has meant in recent years that up to 10 per cent of each age cohort are exempted from service under a broad definition of conscientious objection. The notion of a pluralistic society weakens the very foundation of popular military service.

Comparable trends are at work in the Soviet Union and Eastern Europe, but in vastly different cultural and political settings. Totalitarian control eliminates political and moral debate about conscription. Thus published opinion polls from Poland indicate that young people, while they profess "appropriate" answers to general questions about military service, in specific details reveal strongly negative attitudes toward the realities of conscript service. Only a small minority were positively attracted. Discontent among youth is acknowledged to be widespread in the Soviet Union, and this has its implications for conscript service. In the Soviet military, authorities have to deal with their own forms of social turbulence. They have sought to limit the term of conscripts, have emphasized volunteer recruitment wherever possible, and have closed important branches of the armed forces to all but volunteer personnel. The military has been downgraded as a locus of citizenship training; this function has been transferred to premilitary training in high school, where specially trained military personnel are assigned this task.

In the Soviet Union and the other Warsaw Pact countries the military forces serve as an integral element of the internal security system. These political realities, plus the validity given to the Chinese threat, mean that personal (let alone moral or political) opposition to military service has little or no direct impact on manpower policy. But one should not underestimate the extent to which Soviet authorities are concerned and must take into consideration the attitudes of indifferent youth. In the United States and in NATO, the equivalent opinions exert strong weight in influencing the balance of political decisions about manpower policies.

The introduction of the all-volunteer armed force in the United States means that manpower--as much as or even more than technology--will influence military strategy in the decade of the 1970s. The president's commission in February, 1970 projected an all-volunteer force of approximately 2.6 million, or slightly less than that of the pre-Vietnam buildup. At the time that projection already appeared to be either a major miscalculation or a form of self-deception. In the spring of 1971 civilian officials in the Department of Defense were saying publicly that the post-Selective Service force would be approximately 2.25 million, while privately they indicated a more realistic level of 2.0 million. However, the prospect of a force of 1.75 million is more likely before 1975 and an even smaller force after that date can not be ruled out. The major reduction, of course, will be in the ground forces.

The reduction in manpower partly reflects deliberate national policy, which does not require so large a force. But equally important are economic concerns. There will be intense political pressure to reduce military expenditures below the 1971-72 figure of approximately 8 percent of the gross national product. Any such reduction, however, will be slow and most difficult to achieve, for at any given level of strength, personnel costs in an all-volunteer force will require a markedly greater percentage of the military budget. In the light of the British experience, United States personnel costs can be expected to rise from more than 40 percent to nearly 60 percent of the military budget during this decade. Likewise, unless there is a drastic alteration in weapons procurement policy, the cost of armaments will also rise, since the weapons requested by the military are becoming more and more complex, automated, and costly. Finally, it is highly doubtful that the United States will be able to meet projected recruitment quotas, whatever pay level is offered, without radical internal reorganization of the military such as that described below.

Thus, there are two essential questions: How can the United States forces be redeployed and professionally reorganized so as to articulate with a meaningful and politically responsible foreign policy? How can these all-volunteer forces be recruited, trained, and managed so as to articulate with civilian control and prevent social isolation of the armed forces from the main currents of domestic society? Although there is an element of risk, I firmly believe that a military force of 1.75 million men consuming 8 percent of the gross national product (a percentage that should decline gradually) could support a meaningful defense policy of effective, minimum deterrence rather than a strategy based upon a delicate balance of terror. Such a military force would undertake a variety of national emergency tasks that cannot be performed by civilian organizations, the performance of which would enhance its military effectiveness.

REDEPLOYMENT OF UNITED STATES MILITARY FORCES

The present decade in United States military policy is obviously conditioned by the successes and failures of the past twenty-five years. During these years the United States has pursued one global strategy of nuclear force and two increasingly divergent strategies for its conventional forces-one in Western Europe and one in Southeast Asia. The redeployment of conventional American military forces rests upon recognition of this duality-upon the differences in American interests, responsibility, capacities, and achievements in Western Europe and the Far East. This formulation and distinction does not exclude the problematic issues of the Middle East, Africa, or South America, but rather places them in an appropriate perspective for the purposes of the analysis at hand.

During the past twenty-five years the prospects of nuclear war have been very remote indeed, and this was abundantly clear to detached observers immediately after 1945. There has been a vast amount of literature pointing out that during this period nuclear technology developed a delicate balance of terror; the impact of nuclear warfare would be so destructive and so self-destructive that the results would be of little political advantage to either the United States or the Soviet Union, and therefore the strategy of mutual deterrence emerged. It has been less widely recognized, however, that the absence of a major war is also to be credited to political leadership and to political accommodation in working out arrangements to control the threat of both accidental and premeditated war. The essential political formula was as follows: (1) the Soviet bloc believed that the United States leaders firmly controlled their military establishment and that the United States, on political and moral grounds alone, had ruled out a preemptive nuclear attack on the Soviet Union and (2) the conventional forces of NATO were not to be used to support any movements of national liberation within nations of the eastern bloc.

As the 1960s came to an end, this political formula was strained by the uncertainties introduced by MIRV type weapons and antiballistic missiles. Again, the issue did not involve primarily the technology but the political setting, although these new weapons and counterweapons greatly complicated the search for effective political arrangements. The threat of nuclear confrontation between the United States and the Soviet Union remains remote during the decade of the 1970s as new channels open for mutual political communications and negotiations, both formal and informal. First, the SALT negotiations have become a central forum: at a minimum, prolonged discussion will inhibit the deployment of new weapons, and more effective negotiations will produce ad hoc and partially formalized settlements. Second, partial resolution of the "German question" and unilateral and bilateral troop reductions will serve as a new setting for political discussions and assurances. As China develops her nuclear potential, the United States is required to extend and adapt the political formula of the past twenty-five years to that country; in these circumstances the threat of nuclear war with China will also remain remote.

The problematic issue, therefore, becomes the deployment and redeployment of conventional weapons and troops as adjuncts to the deterrence strategy. In this respect, the experiences and legacies of the past raise sociopolitical questions of importance to any analysis of the probable impact of volunteer forces during the coming decade.

In Western Europe, the stationing of American troops and the system of defense alliances were compatible with European national and political aspirations. Until 1970, stationing the considerable numbers of United States troops required for the NATO strategy of deterrence created relatively few sociopolitical problems at the community level or even that of national sovereignty. In the Far East, too, the stationing of troops and their direct involvement were relatively compatible with local and national aspirations until the end of the Korean War. Progressively, however, the tasks of American troops have become more and more difficult because of the opposition they encounter from national and political forces. Cultural and racial differences between the U.S. and local civilian populations have also been deeply disruptive. Moreover, the fundamental sociopolitical basis of military strategy has been increasingly incompatible with the realities of social change and social structure in Southeast Asia. Finally, there has been the stalemate and atrophying of American military intervention in Southeast Asia due in part to the overestimation by key civilian and military leaders of the impact of conventional strategic air warfare.

In the ever-quickening redeployment and reduction of overseas American military forces, one is struck with the apparent strategic consensus that has emerged; only in Western Europe does there remain a positive function for any significant numbers of United States ground troops. Even in South Korea, the prospect is for an American military presence limited to air and naval units, without ground troops. A contracting system of naval and air bases, selected trained missions, plus an overseas scattering of specialized troops for communications and logistical purposes are the augmenting elements. Thus the question of military posture focuses on the type and number of ground troops required for deterrence and peace keeping through a military presence in Western Europe.

But to identify a common focus does not obscure fundamental differences within the professional military. Military perspectives and doctrines, as held by professional soldiers, have a strong persistence even as they are adjusted to changing technology and an altered sociopolitical environment. The distinction between the "absolutists" and the "pragmatists" still dominates the debate in the day-to-day struggles over military budgets and missions. The trend, because of the wounds to professional pride from the experience of Vietnam, have been more and more toward an absolutist doctrine. This is especially the case in the absence of incisive top military leaders like those who managed the military establishment during World War II and the Korean conflict.

Both absolutists and pragmatists believe that they are sensitive to the interplay of political, economic, and military factors in worldwide international relations. The heavy investment of the military in the politicomilitary education of higher officers has spread a new vocabulary through the ranks of the professional soldiers and the contents of the mass media reinforce this rhetoric. The strongest advocates of each school really live in different worlds--but it should be emphasized that most officers in effect lean one way or the other. The crucial difference lies in the degree to which the professional military man accepts the strategy of deterrence with its implications for the military, and abandons the "killing business" as the organizing principle of his profession.

Deeply held notions in the professional ideology of violence are involved in the debate. At the international level, the absolutists maintain an "assault" perspective even though they are cognizant of the inhibitions that nuclear weapons place on the great powers. They feel that the full political impact of our military forces will be lost unless a "forward" military posture is maintained. Such an assault posture implicitly rejects the formulation of President Nixon's "no more Vietnams" as a passing political slogan. Rather, the deployment of aggressive air and naval patrol forces, forward positioning of bases, and a potential for significant intervention outside Western Europe are required. The strategic concept is in effect a continuation of the notion of the inevitability of armed conflict, but in an altered format. At the personal level the absolutists are fearful that a military without combat experience will decline and atrophy; the models of Sweden and Switzerland are rejected as not professional and--by implication--insufficiently "masculine."

By contrast, the pragmatists, while emphasizing combat readiness, see the military force as playing its role under powerful political and moral constraints. They see the possibility of a successful United States foreign and military policy without an overt assault ideology. For them, the function of the military is the political intent it imparts as much as the sheer destructive power it bears. At the professional level, they are prepared (and even hope) for a career as a soldier without combat, and they believe that combat readiness can be maintained without repressive disciplines and without "satanizing" the enemy.

Each orientation leads to differing military contributions to the persistent questions of force levels, structure, and deployment. For any given budget or force level, the absolutists are at odds with the pragmatists, although officers' attachments to their own services also help mold professional perspectives. The issue, however, is never (for instance) merely "army" self-interest versus an "air force" approach. The absolutist assumes the desirability of maintaining the military budget at the current proportion of the gross national product or increasing it to a higher proportion, whereas the pragmatist is prepared to accept or adjust to a lowering from the contemporary 8 percent. The following excerpts from a hypothetical staff paper on joint planning for a 1.7 million force may afford an insight into contemporary professional debates and struggles.

The absolutist, with an assault concept, would recommend a 200,000-man Marine Corps, equipped with capabilities for tactical nuclear weapons, an Air Force of 500,000, and a Navy of 500,000, which would emphasize attack aircraft carriers to support amphibious or airborne warfare. The ground forces would be limited to 500,000, and most of these would be stationed in the United States, with 50,000 at most in Western Europe. The United States NATO ground troops would be part of a fully automated battlefield with electronic surveillance and highly "modernized" weapons and deployment, and they would operate with tactical nuclear weapons as a trip wire. In addition, 15,000 of the ground troops would be a special force trained for armed reconnaissance and counterinsurgency operations behind enemy lines.

By contrast, the pragmatist would limit the Marines to 100,000, specify smaller naval and air forces of approximately 450,000 each, with naval emphasis upon antisubmarine warfare. Some 700,000 men would be allocated for the ground force, of which 150,000 would be stationed in Western Europe (representing half the 1971 force level), and they would be closely articulated with emerging militia systems of Western European nations. The emphasis on special forces would be limited to a few thousand specialists. The forces stationed in the United States would be heavily involved in national emergency work as well.

RECONSTRUCTING THE ARMED FORCES

The adaptation of the United States armed forces to the end of conscription means that operational procedures that have grown up since the end of World War II require comprehensive review. The armed forces are experiencing a deep "generation gap," in that the cadres of middle-level officers are better prepared for institutional change than older and higher-ranking personnel although, of course, there are important exceptions. The tension is strongest in the ground forces, which have had to bear the burden of Vietnam and for whom the transition to an all-volunteer force is the most difficult.

An armed force which is smaller, recruited on an all-volunteer basis, and organized more and more as a force in being reverses or at least halts the historic trend of the United States military establishment toward civilianization. South Vietnam drastically delayed the emergence of a new organizational format; in fact, military historians will look back on the Vietnam war as one that extended the life of traditional World War II perspectives and strengthened barriers to change. The transition to an all-volunteer system therefore will take place abruptly and under highly unfavorable circumstances.

In the American environment, it was to be expected that economic incentives and financial rewards would be the main thrust for shifting to an all-volunteer system. In the fall of 1971 Congress passed a comprehensive military bill that raised the base pay of an enlisted recruit to over $260.00 a month. Base pay plus allowances and fringe benefits voted at the same time will bring the annual income of an army private close to the very symbolic figure of $5,000.00. The additional costs of these 1971 increases totaled over 2.5 billion dollars. This figure comes close to the total cost increase for personnel which the President's Commission on an All-Volunteer Force estimated would be required to create an all-volunteer force. In short, the economic incentives approach has been put into operation, but there is every reason to believe that major additional increases will have to be instituted in the future.

Meanwhile, other approaches tend to be neglected. Congress dismissed without any debate a system of volunteer national service which would allow young people to select between military and civilian service. Such a program might have strengthened the social definition of service to the nation, and this in turn would have greatly facilitated recruitment by creating a new legitimacy for governmental, community, and military service. Moreover, the needs for basic policy changes in recruitment, career lines, education, and deployment are being faced only very slowly. New operating procedures, especially in the ground forces, have been limited to improving the physical character of barracks for enlisted personnel, new recreational resources, wider freedom in personal appearance, and modification of some aspects of the daily routine of garrison life, particularly early morning reveille. Skeptics have described these steps as a "cosmetic" approach to the problem. Likewise, there is a powerful reluctance to explore options for fundamental "institution building" in the armed forces; it is as if to do so would imply a defect in earlier practices. Nevertheless, some areas of change do hold promise of creating volunteer forces appropriate for a democratic society and a military posture of effective deterrence.

First, an all-volunteer service requires a fundamental redefinition of the tenure of a military career in order to strengthen its civil service basis. A significant proportion of both officers and enlisted men will continue to serve for six years or less; for them, military experience is an interlude in an essentially civilian existence. This type of military service appears to be viable, although there are many attendant issues to resolve.

Another group, however, will serve in the armed forces (as now) for much longer periods of time, often up to twenty years. For this group military service must be redefined as one step in a two-step career that is, a lifetime career in the public service with a military assignment as the first portion, and a civil service assignment the second.

For enlisted personnel successful completion of a specified period of service, such three periods of enlistment, could constitute effective entrance into civil service employment. The United States Civil Service, under the Department of Labor, would have the responsibility for placing the individual in the federal service or, by negotiation, in state or local governmental service. Such a career system would broaden the basis of recruitment, attract personnel of appropriate quality and eliminate the costly system of reenlistment bonuses and existing pension plans. If an enlisted man transferred to the civil service establishment he would take with him pension benefits equivalent to those of civil employment, and these would be paid to him on retirement.

An equivalent system would operate for officers but would go into effect only after the size of the officer corps had been reduced. In addition, the length of the term of service for officers would be made more flexible. Exit with appropriate pension benefits after ten to twelve years is essential to have a flow of personnel which will integrate with the rank structure and the military tasks to be performed.

Second, the shift over to an all-volunteer system must deal with the inflated rank structure, which is both wasteful and keeps younger men from obtaining meaningful assignments. Likewise, the services must face the fact that too many general officers have accumulated in the three services. The excessive number of such officers thwarts the assignment to important posts of younger men prepared to adapt to the changing environment. The army, in particular, has a deep division between the junior and mid-career officers who actually fought in South Vietnam and the ranking personnel who flew over the battlefield or were in top command positions. The rapid incorporation of men in their forties into the general officer group is essential to heal the breach and to offer an incentive for able mid-career officers to remain in service. To deal with this problem, generals will have to be retired at a rate faster than normal until their numbers have been significantly reduced. Needless to say, such an objective will be difficult to attain.

Third, the existing worldwide personnel system which leads to continuous, excessive, expensive, and disruptive rotation can no longer be justified. Instead, the armed forces, and particularly the ground forces, will have to develop a modern version of the British regimental system--or in the present context, a modified brigade system. Each man would have a basic unit and a significant portion of his military career would be spent within that brigade. For the Navy, a home port concept and for the Air Force a home basis would serve as the equivalent.

Fourth, the armed services will have to recognize that underemployment is a powerful source of negative attitudes toward a military career, especially among young officers. In the past, military personnel were less sensitive to the stimulus and responsibility of their initial assignment. They assumed that war would "break out" at some future time, and then they would be fully engaged. The heavy reliance of the military on short-term officers and the emergence of the strategy of deterrence make the issue of the intrinsic relevance of the day-to-day job and the avoidance of boredom and a sense of futility very important. Fundamental changes are required in military training so that many training functions may be transferred from specialized and centralized units to operational units. This is needed both to improve training and to reduce the amount of underemployment in operational units.

Fifth, the military services will have to place a stronger emphasis on officer candidate schools for recruitment and training of new officers. The end of conscription will tend to reduce the pool from which qualified officers can be selected. Officer candidates will become less socially representative--they will be predominantly from the south and the southwest and from rural and small-town areas. ROTC units will have to be reorganized so that any college student in the United States either on entrance into college or when he becomes a junior would have access to a collegiate ROTC program. In each of the ten major metropolitan areas there should also be a composite program administered by an existing ROTC group which would enroll students from any accredited college in the metropolitan area.

Sixth, the present system of in-service professional schools needs to be consolidated. The present system is wasteful, repetitious, time-consuming, and often merely mechanical. Military officers require extensive education, but many competent officers consider the present system an excessive diversion from professional service. The system should be reduced to a two-tier system, with the interservice component distributed to the service war colleges. A strong emphasis on brief courses to handle new developments in organization and doctrine would be desirable, as well as permission to substitute civilian schooling for attendance at advanced military schools.

Seventh, the academy programs should permit a one-year

attendance at a civilian university, for example, in the junior year. Alternatively the academy program could be a five-year program with one year free for civilian work experience or service in the enlisted ranks.

Eighth, the services must establish a Department of Defense commission to revise the essentials of discipline. Combat-ready forces, fully sensitive to their heroic traditions and under the closest operational control, can be trained and maintained without brutality, personal degradation, or "Mickey Mouse" discipline. The United States Marine Corps may be permitted and able to maintain, as its top commanders insist, its traditional organizational code of repressive basic training, but an all-volunteer military force must face openly and candidly questions of authority and military forms. For example, extensive saluting on military bases serves no purpose but to degrade the act; saluting can have meaning as a part of crucial and selected formations.

Ninth, military justice is being transformed by civilian court decisions and will emerge closer to civilian procedures for nonmilitary offenses.

Tenth, there exists a good deal of concern that under an all-volunteer system, the physical concentration of military families on bases will contribute to their social isolation. Moreover, the current trend toward more off-base housing will not necessarily be a positive factor here, since relocation in a civilian community does not automatically produce social integration into the larger society. Where feasible, military families should have an element of choice in their housing, since for most of them access to the facilities of the military base is essential to meet the pressures that the military places on family life.

Instead, the quality of integration into civilian society depends upon personal initiative and membership in voluntary religious and community associations, as well as upon military regulations and concepts about civic participation. In Germany, the idea of the "civilian in uniform" has been pressed to the point where regular military personnel--both officers and enlisted men--are permitted to stand for political election while on active duty. In the American context, the need to maintain a nonpartisan (that is, nonparty) affiliation remains essential, but a broader perspective on civic participation is possible. Military personnel should be permitted to serve on local school boards, run in nonpartisan local elections and be members of government advisory boards and public panels when they have qualifications and interests.

It is not the responsibility of military personnel to defend and publicize official military policies; this is the task of elected officials. But the military are not hired mercenaries; they cannot be mechanically deprived of participation in community and public affairs. By law, and particularly by judicial decree, military personnel are exercising their particular forms of free speech and citizen petition. The prospect of trade unions, without the right to strike, is a real possibility in the military. This can be done without interfering with professional responsibilities. In a truly pluralistic society, with dignity and good taste, military personnel, while on active duty, should be able to attend education, community, and public affairs meetings and assemblies and state their views on the legitimacy of their profession.

REDEPLOYMENT OF FORCES

The all-volunteer armed force faces a deep dilemma in the subsequent steps of its strategic redeployment. On the one hand there is the powerful self-fulfilling prophecy which is already at work: each reduction in force serves only to dampen new recruitment, especially officer recruitment. (The massive survey of the British forces carried out in 1969, nine years after the end of conscription, found that concern with future force reductions was the major source of professional discontent and one of the main reasons for planning to leave the forces.) Why enter a profession whose career and promotion opportunities are highly uncertain and declining? The plan for a civil service base for the military profession, as described above, is one device for handling this problem. Therefore, paradoxically, the faster the initial reduction to a long-run troop level, the more readily the adaptation can be made to a volunteer force. The phased withdrawal of troops from South Vietnam from a high point of over 500,000 in 1968 represents the largest single component in the overall reduction of manpower. The next step, and especially in ground troop reductions, will have to come from NATO forces. This therefore, is the other side of the dilemma: the faster the reduction the greater the political difficulties of adapting Western European defense policy to new realities.

Although plans are projected for negotiations with the Soviet Union on European security and on mutual and balanced reduction of forces, economic pressures have committed the United States to unilateral reduction of United States contingents to NATO. Senator Mike Mansfield has been a persistent advocate of troop reductions in NATO, and there is every reason to believe that President Nixon accepted some direct reduction in return for support from members of the Democratic party for his economic program of 14 August 1971. A reduction of 5 to 10 percent of the 310,000 troops could be made immediately without any diminution of military effectiveness. The United States forces in Europe have become excessively bureaucratized, and such a reduction would in fact serve to increase operational morale and reduce the tensions associated with boredom.

Assuming a transition to an all-volunteer force and accepting the goal of a reduction of United States troops in Western Europe to 150,000 in three to five years, the United States must (1) seek effective negotiations with the Soviet Union and (2) initiate a political strategy of reratifying the basic principles of NATO in the contemporary context. If troop reductions, no matter how limited, are seen as first steps toward a United States neo-isolationism and a withdrawal from Western Europe, the Brandt initiatives will collapse, and there will be a major political crisis in Western Europe, with profound implications for the United States. The actual size of our troop commitment is not more important than the stability of our intention.

Only ratification of a new NATO treaty and simultaneous negotiation with the Soviet Union will suffice. This can be accomplished by dispatching a United States delegation of the highest level--including bipartisan representation from the Senate and House of Representatives--to prepare a new treaty for ratification by the president and the Senate. The essential element would be the restatement of long-term United States troop commitment in Western Europe. Such a declaration would set the minimum force level under various conditions, including successful negotiations at the NATO-Warsaw Pact conference.

Such a reratification of NATO principles would include the following developments for NATO and United States forces. First, the United States could establish the fact that the projected manpower reductions in Europe were linked to an increase in the United States strategic reserve available for airborne redeployment to Europe should the future international situation require it. The Nixon doctrine of "no more Vietnams" implies that NATO requirements will have higher priority on this strategic reserve, which should help reassure Western Europe. Second, a NATO rear headquarters in the United States would be in order. At this headquarters, European NATO officers could be assigned for planning and staff work and for direct command with United States strategic reserve forces. Third, it should be possible to alter United States planning and commitment from a year-to-year basis for NATO to a five-year basis. For example, it would be useful to explore the establishment of a NATO mutual security bank into which the United States would make payments as a sign of future commitments. Fourth, the United States and NATO nations should encourage new manpower systems in Western Europe. It makes little or no political sense for the United States to urge Western Europe to maintain a traditional concept of conscription at a time when it is moving to an all-volunteer force. For Europe, and especially for West Germany, militia systems, including six-month conscript service, need to be developed. For the United States, in addition to the reforms mentioned above, the United States reserve forces need to be fundamentally reorganized into three elements. One major part should be a ready reserve, capable of two-week deployment. Another part would be made up of individuals who, following the Israeli pattern, would serve briefly each year as filler personnel in operational units. A third part might be the traditional inactive reserve--but it would function as a real manpower pool, receiving some limited training and being compensated accordingly.

Fifth, the following guidelines are proposed for the NATO-Warsaw Pact conference of mutual force reduction.

1. The immediate establishment of a hot line between the headquarters of the Warsaw Pact and NATO headquarters and a joint Warsaw Pact-NATO liaison staff for information and communication purposes. These measures would reduce the threat of accidental war and implement the surveillance aspects of mutual security arrangements. The joint NATO-Warsaw Pact liaison staff would be located in neutral Switzerland.

2. A step-by-step negotiation first to a 25 percent and then to a 50 percent balanced reduction of the level of ground forces,

taking into consideration the strategic positions, weapons balance, and lines of communication of both NATO and the Warsaw Pact nations.

3. The establishment of an effective system of mutual surveillance both to guarantee compliance with negotiated terms and to monitor deterrence capabilities. On-the-spot inspection is probably not necessary, if aerial and electronic surveillance is organized on a joint basis.

All these changes in military organization and troop deployment involve shifts in professional ideology and self-conception. It is now more than a decade ago that I offered in The Professional Soldier, in assessing alternative futures for the military, a definition of a constabulary force.* The constabulary concept provides a "continuity with past military experiences and traditions, but it offers a basis for radical adaptation of the profession. The military establishment becomes a constabulary force when it is continuously prepared to act, committed to the minimum use of force, and seeks viable international relations, rather than victory, because it has incorporated a protective military posture doctrine."

In the prologue to the new 1971 edition of The Professional Soldier, I underline the conclusion that prolonged hostilities in Vietnam have unfortunately diverted the attention and energy of the military from such a goal. A mass of materials have been written about the changing role of the military in contemporary society, the essence of which is widely debated by military officers. The military profession is divided and indecisive about how much of the emerging doctrines it will accept. The notion that the military is mainly in the "killing business" dies slowly. But the vitality of the military depends on the transformation of its self-definition to one in which peace keeping is its legitimate role.

Moreover, it does not appear that the military can renew its vitality unless it comes to see itself employing its facilities in a wide range of national emergency functions. The basic issue is not, as traditionalists hold, that the military should not be diverted from its fundamental mission. The military have long engaged in national emergency functions. But the nature and contents of these functions must change. In reconstructing the military, it is essential to make effective use of its manpower and vast resources to keep it an active and responsible institution. The notion "deterrence is not enough," does not necessarily imply an assault mentality. It is also an outlook required to attract and retain bright and highly motivated men who wish to avoid underemployment and get on with the job of social change.

Clearly, the military cannot engage in activities or programs which are better performed by civilian agencies. The essential issue is to make use of its standby resources that is, its ability to respond to emergencies, broadly defined, and to improvise in a nonroutine fashion. The military are already deeply involved in control of the effects of natural disaster--floods, hurricanes, and the like pose emergency situations that require their flexible resources. To natural disasters can be added the increasing scope of man-made disasters; oil spills, power failures, and chemical and atomic accidents are likely to increase rather than decrease. The armed forces are indispensable in a vast array of air and sea rescue work, to which is being added, on an experimental basis, medical evacuation, especially of victims of road accidents, where alternative facilities are not available.

But the major frontier rests in the arena of environmental control, and the handling of particular aspects of pollution and destruction of resources. The Corps of Engineers has moved in this direction, but only the first steps have been taken. Many units in the armed forces have contributions to make and the notion of a military career as part of a civil service career means new patterns of assignment between military and civilian agencies. An armed force of over one and one-half million men offers a significant manpower pool, and one that is urgently needed given the economic pressures of contemporary American society.

The concept of the all-volunteer force does not deny or destroy the difference between the military and the civilian, for to do so runs the risk of creating new forms of tension and unanticipated militarism. On the contrary, it calls for a special sensitivity to the distinction between the civilian and the military, each with its specific responsibilities. But the boundaries between the civilian and the military can be redefined without excessive civilianization of the military. It is not a force composed of men who, on the average, will spend the bulk of their working lives in the military; projections of a highly stable force forecast that the length of service for enlisted men in the army will be no more than five years, with an equivalent figure for officers. In fact, a military which is engaged primarily in deterrence does not have to be a profession with less, but can be one with more civilian contacts.

The all-volunteer armed force marks the end of the historical era of the mass armed force. The rise of the mass armed force was not purely a military development but reflected the sociopolitical trends of nationalism. The present internal tensions and the crisis in legitimacy within the armed forces have meant that the decline of mass conscription, although delayed by the war in Vietnam, will take place rapidly and without significant resistance in American society. The form and character of the all-volunteer force, again, will not be a purely military phenomenon but will reflect the character of the larger society. The all-volunteer armed force will be rooted especially in those elements of American society which continue to be the carriers of traditional nationalism. But the military can also reflect and incorporate new forms of transnationalism which already exist both in its own ranks and in civilian society. Under these circumstances, it will be the duty of the civilian society to assume an active role in directing the military to redefine its professional perspectives and to help it understand that peace-keeping through a military presence, deterrence, and participation in the control of national emergencies are the modern definitions of the heroic role.

NOTE

*The Professional Soldier: A Social and Political Portrait. New York: The Free Press, 1960; revised 1971.

ALTERNATIVE STRATEGIES AND BUDGETS FOR MILITARY SECURITY

Seymour Melman

In the foreseeable future, plausible military security requirements of the United States will require the operation of armed forces. This requirement will surely continue until there is major international agreement and implementation of disarmament.

This paper defines the main parameters of U.S. military security forces, designed to afford competent protection for the people of the United States against military aggression from without, and to insure capability for American participation in international peacekeeping.

The main approach here is not cost but competence in a military security force. Competence and adequacy in a U.S. military security force are defined in terms of:

(1) appropriateness to military security goals;

(2) appropriateness to present and predictable technology;

(3) appropriateness to basic security requirements of the United States.

This analysis assumes that the security of a society consists of protection against destruction from without, as well as the well-being of its people. A competent security policy must serve <u>both</u> these ends. The developments that led to the present analysis include emphasis on military forces in the federal budget to a degree that has caused depletion in the quality of life for the American people. By spending more than $1,100,000,000,000 on armed forces from 1946 to 1970, the American people have been deprived of at least that much productive goods and services, leading to such deterioration and turbulence as to cause sensible men and women to doubt the viability of American society.

The consequence of such reasoning is this: Appropriateness of armed forces to the general security of a society means that methods which would be self-defeating to either side of security are thereby ruled out. During the last decade, the sustained priority given to the design and operation of military forces with open-ended security goals has produced unavailability of capital and skilled manpower for many civilian activities and industries. As a result, many facets of American life have become clearly depleted and the quality of life finds many Americans at a serious disadvantage industrially, technically and economically.

For example: The American housing industry has clearly failed the nation. The supply of medical services has become inadequate to the crisis-producing point. An industry as fundamental as the one that produces electric power has been failing the nation, with recurring brownouts in the northeastern states during the summer of 1970. The steel industry of the United States is becoming progressively less competent to hold the domestic market against the competition of foreign producers. Informed persons in the electronics industry advise that the Japanese electronics industry is likely to become pre-eminent in the world civilian electronics market by 1973.

Military security policies that require open-ended military budgets produce effects that are self-defeating with respect to the security of the United States as a whole. Accordingly, such policies and the methods specific to them are ruled out as inappropriate for this definition of an adequate military security policy for the United States.

Until recently, the operative military security goals of the United States included the requirement that armed forces be capable of fighting three military conflicts at once:

(1) a war in Europe, presumably nuclear with the U.S.S.R.;

(2) a Southeast Asia war;

(3) a lesser military engagement in Latin America.

In addition, military forces have been designed for open-ended capability of "flexible response" in diverse situations. These requirements meant that U.S. armed forces had to have in hand everything from a pocket knife to a twenty-megaton hydrogen bomb warhead, in diverse delivery systems. These military security requirements of the last decade should be discarded in favor of revised military security goals.

Appropriate military security objectives for the United States should include, for the foreseeable future, armed forces that are competent to:

(1) constitute a credible deterrent against nuclear attack;

(2) afford clear capability for guarding the shores of the United States; and

(3) give the United States substantial competence for participating in international peacekeeping operations with other nations.

For these purposes, armed forces should use present and predictable technology, making use of an available array of military options for the design and equipment of forces. Appropriate technology, however, excludes various available or conceivable technical options: it excludes the buildup of nuclear overkill forces as technically and humanly inappropriate; it excludes the manufacture and storage of large-caliber naval weapons and shells, since such weapons have been superseded; it excludes systems which would not be effective in the event of a Soviet first-strike; it excludes chemical and biological warfare weapons which are not only abhorrent to most of mankind, but have the technical possibility of backfiring upon their user; it excludes systems like the ABM which do not give a physical shielding capability while opening the wielder of ABM forces to serious misinterpretation of intention by an opponent.

The military force to serve the function of strategic deterrence for the United States should consist, preferably, of the Polaris submarine fleet with its complement of missiles. This delivery system has capabilities that make it a preferred instrument for this military security function, for the foreseeable future. These desirable characteristics include: reliability of central control; variation in the position of these vessels; difficulty of detection; capability for rapid movement; long duration cruising capability; reliability of control at each location; reliability of missile performance.

The Polaris fleet of forty-one nuclear-powered submarines with sixteen missiles per vessel can deliver 656 nuclear warheads, without any multiplication by the MIRV technology. This number of warheads should be reckoned against the fact that the population-industrial system of the U.S.S.R., for example, is concentrated in 156 cities of 100,000 persons and over. Obviously the Polaris system alone, in 1970, includes multiple overkill capability vis-a-vis the U.S.S.R. Early warning is by satellite system.

For strategic forces, it must be emphasized that the requirement can only be adequacy and not superiority. Adequacy means plausible competence to perform a particular function to the degree that this competence is well-known and therefore comprises a credible threat to a possible opponent. Superiority in the strategic realm is not attainable. A multiplication of nuclear warheads and their delivery systems produces capability for theoretically multiplied overkill, which is not militarily or humanly meaningful. Therefore, a comparison of national strategic forces in terms of numbers of deliverable nuclear warheads is not a relevant comparison in the nuclear era.

Armed forces to guard the shores of the United States in a competent manner should include combined land, sea and air units. Seaborne units include four aircraft carriers and their escort vessels, two carriers on each coast. These vessels, nuclear-powered, have long cruising range and high speed cruise capability. The relevant air forces also include land-based aircraft on coastal patrol. The land forces for this function include a portion of the 100-battalion fighting force that is defined below.

Capability of U.S. armed forces for participating in international peacekeeping, and for guarding the shores of the United States as well, can be adequately provided by a force of 100 airborne battalions, each consisting of about 400 men. These airborne units, with integral transport and support capability, consist of highly-trained soldiers armed primarily with high-firepower lightweight weapons and capable of great mobility on the ground and in the air.

It is emphasized that this conception of military security forces for the United States excludes an effort to define forces for either first-strike nuclear wars or for a succession of Vietnam-type wars. The exclusion of nuclear first-strikes as a military security goal permits substantial reduction in strategic forces. Elimination of a succession of Vietnam-type wars as a security goal makes possible substantial reduction and change in the character of land-based naval and air forces. The massive complement of eighteen divisions in the U.S. Army can be justified on the grounds of preparation to either fight World War II a second time or to sustain a continuing program of wars of intervention and occupation. The elimination of these open-ended military security objectives makes possible a substantial redefinition of force requirements.

The following is a preliminary estimate of manpower required for direct operation of the forces defined above. These estimates are based upon a tentative view of such forces, and have been compiled without benefit of access to the sorts of detailed forces data that are normally available within the Department of Defense. Nevertheless, the following estimates are given here insofar as they suggest an order of magnitude of force requirements that is strikingly different from those prevailing at the present time.

	Estimated Direct Manning Needs
Strategic deterrence force	
41 Polaris submarines, double crew	28,700
Guarding the shores of the U.S.	
4 nuclear-power carriers, with 6 destroyers/carrier	21,800
20 mine countermeasures ships	1,000
coastal air patrol and tactical, 100 planes, land-based	3,000
Airborne combat units (battalions) for international peacekeeping	
100 combat units - 40,000 men	
tactical air-ground support - 16,000	
air transport - 10,000	66,000
TOTAL	120,500

The men operating these units are backed up by a complement of support forces in tha ratio of seven to one. By this reckoning, the support forces would comprise 843,500 men.

Altogether, the uniformed forces required to conduct the operation of the military security force suggested here comprise 964,000 men. This contrasts with the present armed forces including about 3,200,000 uniformed men and women.

It is appreciated that the foregoing design of armed forces requires a transition period. What is crucial is the definition of objectives at the earliest time and the direction of motion toward those objectives.

In this part of the paper, I will suggest a redesign of U.S. military forces based on a series of estimates (incomplete) of possible reductions in the present armed forces budgets, taking into account the sort of major revision in forces that is suggested here. It is emphasized that the following estimates of budget reductions refer to only part of the present U.S. military establishment. These estimates are included here, despite these limitations, for they suggest the operational meaning of cost changes owing to changes in military security requirements and in forces adequate thereto.

In sum: I estimate that strategic weapons, plus their elaborate warning systems, will require, directly and indirectly, about 450,000 uniformed military personnel in 1972. The cost of the administration's Strategic Forces program will be about

$16.3 billion, or 21 per cent of the proposed defense budget (FY 1971).

The recommended deterrent force, based on the Polaris fleet, with full supporting staffs and services (research, medical, training, administrative, etc.), would cost about $3 billion per year, compared with the present $16.3 billion price tag on the administration's expanding Strategic Forces.

The cost of operating and further developing the General Purpose Forces in 1972, together with supporting civilian employees and supplies, would be $59.5 billion, or 75 per cent of the administration's defense budget. This includes about $6 billion for Vietnam operations.

My proposal for a U.S. military security policy includes guarding the United States and participating in international peacekeeping. These functions would be performed by combined air, sea and land forces.

U.S. defense and peacekeeping forces of this kind could be operated by 735,000 men as compared with the 2,100,000-man level programmed by the Nixon administration for 1972. I am unable to give a separately itemized estimate of the cost of operating the force I propose since it includes military units of a type not presently operated. A rough approximation is possible, however. Present forces cost about $30,000 per man-year (including new weapons, etc.); on this basis, the cost of the proposed non-nuclear defense and peacekeeping force would be 735,000 men multiplied by $30,000, or $22,050,000,000.

The Nixon administration proposes a 1972 budget of $79.2 billion for the Department of Defense. I suggest that the military security of the United States can be competently served with armed forces costing two-thirds less, between $25 and $29 billion. The difference, approximately $50 billion per year, is what is needed to start restoring the United States economically and morally.

NAVAL FORCES

The Attack and ASW Carriers

The Enterprise and the three Nimitz class CVAN's (nuclear attack carriers) would form the basis for three attack carrier task groups composed solely of nuclear-powered ships. These groups would thus be able to operate at high speeds for prolonged periods of time while remaining free of the costly and vulnerable supply trains required by conventionally-powered ships.

Estimated annual savings: $316.25 million. [1]

The Forrestal class carriers would all be retired.

Estimated annual savings: $4,400.00 million. [2]

The Essex and Ticonderoga class ASW carriers would all be

deactivated, as it is impossible to justify any further retention of these aged and impotent ships in the face of the new developments in submarine technology.

Estimated annual savings: $1,100.00 million.[3]

The Escort Forces and the Cruisers

The DX (escort destroyer) and DXG (guided missile escort destroyer) would be immediately curtailed. The Navy would only build long-core-life nuclear-powered frigates, with AAW (anti-aircraft warfare) and ASW capabilities, for the three attack carrier groups.

Estimated immediate savings: $3,796.00 million.[4]

No more conversions would be made on existing escort vessels and only the most effective conventional ships would escort the carriers during and after the transitional phase.

Estimated annual savings: $140.00 million.[5]

The active force level of 239 escort ships would be cut to 48 of the newest and most capable ships in the escort fleet, to be used in rotation with 24 at sea and another 24 kept operational.

Estimated annual savings: $2,570.00 million.[6]

All the gun cruisers would be deactivated, as their mission would not be in keeping with our military goals. The light guided-missile cruisers would be replaced by missile frigates. (Thus we could eliminate the flagship scenario.) The Long Beach would be incorporated into an attack carrier group.

Estimated annual savings: $220.00 million.[7]

There would, then, be no further purchases of any large-caliber ammunition for these ships.

Estimated annual savings: $125.00 million.[8]

The Amphibious Forces

Work on the LHA (amphibious storage) type ships would stop immediately. These are merely the FDL's (fast deployment logistics ships) for the Marine Corps under a different name. They would not be needed for meeting future military objectives. Their cost has risen to $185 million per ship.

Estimated immediate savings: $313.50 million.[9]

Estimated total program savings: $1,370.00 million.

The 89 ships in the amphibious assault forces would be immediately deactivated. These ships average 19 years of age and are not designed to accommodate the changes in the technology of the 1970's. Their mission would not be in keeping with the military objectives we have set down.

Estimated annual savings: $200.00 million.[10]

LST (landing support tank) production, Regulus submarine conversion, and LPS (amphibious fire support vessel) contract formulation would all be immediately stopped. Large, slow amphibious vessels would be inconsistent with our proposed military objectives.

Estimated savings: unknown.

Mine Countermeasures Ships

There would be a reduction from 73 ships to 20 with a larger dependence upon the new helicopter techniques. All the ships retained would have to be well under a preset age limit of 15 years and would have to be constantly re-evaluated in order to maintain a force which could meet the problems posed by modern technology.

Estimated annual savings: $100.00 million.[11]

The order for 16 MSO (ocean-going minesweeper) types would be cut in half immediately. The 8 remaining ships would be retained as part of the 20 mine countermeasures craft. Such ships would no longer be needed to open large beach-heads for amphibious assault and support.

Estimated immediate savings: $75.00 million.[12]

The MCS (mine countermeasures support ship) development program would likewise be re-evaluated to prevent a repetition of the Ozarks fiasco.

The MSO conversion program will be curtailed. The conversion process has risen in cost from 4.8 million per ship to 5.2 million, and the time required has doubled.

Estimated immediate savings: $22.40 million.[13]

Fleet Auxiliary Ships

The 373 ships now active would be retired. The auxiliary function can be performed by commercial ships with built-in naval servicing capability -- e.g. high-line transfer and underway refueling. Note that nuclear propulsion of major ships means limited requirements of this sort.

Estimated annual savings: $700.00 million.[14]

The AOE (fleet oiler and ammunition ships) and AE (ammunition ships) building program would be stopped, as the force levels of these ships are already out of line with our new fleet designs.

Estimated immediate savings: $400.00 million.[15]

The Submarines and Related Equipment

The 58 fleet-type submarines would all be re-evaluated to determine their true value as part of the system to insure the survival of the Poseidon submarines. Those which are of a high-cost/low-effectiveness nature would be retired, and the savings would be put toward the 668 class of SSN's (nuclear attack submarines).

Funds allocated to the Mark 48-1 and -2 torpedoes would be curtailed, as these are for use against surface ships and an unnecessary redundancy in our arsenal.

Estimated immediate savings: $40.00 million.[16]

The present U.S. stockpile of Polaris and Poseidon missiles

possesses sufficient overkill capabilities to make further procurement of Poseidon missiles totally irrational. This is especially true if the missiles are equipped with MIRV warheads. The small amount of accuracy gained by using Poseidon instead of Polaris does not justify the conversion of submarines to Poseidon missiles. The program should be halted.

Estimated annual savings: $1,044.30 million.

Estimated total program savings: $5,500.00 million.[17]

The ULMS (undersea long-range missile system) duplicates the purpose of the Poseidon missile. The additional range gained by using the ULMS does not make it more of a deterrent than the Poseidon. The deployment of ULMS would require conversion of submarines from Poseidon to ULMS when they are just now being converted from Polaris to Poseidon. Therefore, the further development of ULMS would be unjustified and should be halted.

Estimated annual savings: $44.00 million.[18]

SABMIS (sea-launched antiballistic missile intercept system) is a water-based ABM system. Its effectiveness would be in just as much doubt as the land-based ABM and should be eliminated.

Estimated annual savings: $3.00 million.[19]

Pilots and Training

Pilots account for only 2.5% of all military personnel, while their training costs are 1/3 of the total spent for training. The reduction in carrier forces would cut Navy pilot levels down by more than 2/3.

Estimated annual savings: unknown.

The program for constant retraining of all pilots and naval flight officers of the rank of commander and below would be ended, as the demand for pilots would decrease.

Estimated annual savings: unknown.

The Lexington would be released from her pilot training role and deactivated. Her duties would be assumed by a Forrestal class carrier.

Estimated annual savings: $40.00 million. [20]

The Planes

As the CVAN force proposed would be a drastic reduction from the 15 currently envisioned, the number of planes required would decrease proportionally. The 38 F-14 aircraft have already cost $760 million. This force level should be evaluated as the 4 CVAN's would be the only ships carrying these planes.

Estimated annual savings: unknown. [21]

The Phoenix missiles, designed for use on the F-14 aircraft, could also be reduced at a savings of $1.3 million per missile to meet the new requirements.

Estimated annual savings: $105.10 million. [22]

The A-7 procurement should be stopped immediately, as the

4 carriers would require fewer than the number of planes already available.

Estimated immediate savings: unknown.

The Marine Corps would no longer require the Hawker Harrier program.

Estimated immediate savings: $118.30+ million.[23]

The EAGB and KA-6D (electronic countermeasures configuration and tanker for the A-6 Intruder) would be cut back to put them in line with proposed force requirements.

Estimated annual savings: unknown.

The P-3C (ASW plane) program would be speeded up, and the Sonobuoy procurement would go on as scheduled.

STRATEGIC (NUCLEAR) FORCES OTHER THAN NAVAL

Minuteman

The Minuteman missile system is a redundancy on the Poseidon system. Minuteman sites are fixed and considerations of their vulnerability led to the development of the Polaris and Poseidon missiles. The further deployment of Minuteman missiles would be terminated.

Estimated annual savings: $686.00 million.[24]

Safeguard Anti-Ballistic Missile System

The purpose of ABM is to protect U.S. Minuteman missile

sites from attack by Soviet SS-9 missiles. However, the Poseidon missiles were also deployed to hedge against the SS-9. This duplication of weapons is considered wasteful and unnecessary. If further deployment of Minuteman missiles were to be halted as strategically useless, it would certainly be unnecessary to deploy an ABM system of dubious value for the protection of existing Minuteman sites.

In addition, the effectiveness of ABM is doubtful (radar could not track a low-flying attack) and it can never be tested due to the ban on testing nuclear weapons in the atmosphere. Such a test, involving airborne nuclear-armed missiles, would risk the destruction of population areas as well as the release of nuclear contaminants into the atmosphere. Therefore, the deployment of ABM should be immediately terminated.

Estimated annual savings: $1,200.00 million.

SAGE Bomber Defense System

SAGE (Semi-Automatic Ground Environment) Bomber Defense System is an obsolete air defense program. It is ineffective against low-altitude bomber attacks. The Soviets do not have a sizeable intercontinental bomber threat. If SAGE were made purely a warning system rather than a full defensive system, a savings of $600 million per year would be possible. However, it still could not track low-flying bomber attacks and would be incapable of tracking missiles while itself being vulnerable to missile attack. Therefore, its effectiveness even as a warning system would be negated. The entire SAGE program should be eliminated.

Estimated annual savings: $1,340.00 million.

Estimated total savings: $5,000.00 million.[25]

(Note: All further estimated annual savings figures are for Fiscal Year 1971.)

Multiple Independently-targetable Re-entry Vehicles

MIRV can only add to overkill and accelerate the arms race.

Estimated total savings: unknown.

FB-111 Strategic Bomber

The doubt as to the effectiveness of the bomber weapon does not warrant the expenditure of millions of dollars, especially when it does not offer any strategic advantage over existing weapons.

Estimated annual savings: $84.90 million.

Estimated total savings: $1,200.00 million.[26]

B-52 Bombers

The defense and retaliatory-strike capabilities of bombers is highly questionable in the face of advanced missile attack and defense systems. As a result, its effect in a nuclear war would be extremely small. The B-52 is no longer an acceptable weapon and should be eliminated from the strategic forces.

Estimated total savings: $1,500.00 million.[27]

Advanced Manned Strategic Aircraft (AMSA, B-1)

In 1964, Robert McNamara stated to Senator Ford that a very, very low percentage of the force that destroys the Soviet Union would be delivered by bombers. The AMSA can just as easily be shot down by defense missiles as the B-52 which they are scheduled to replace. Therefore, the effectiveness of the AMSA becomes highly doubtful and would serve only to add to the already massive overkill capability of U.S. forces. The project should be terminated.

Estimated annual savings: $100.00 million.

Estimated total savings: $8,800.00 million.[28]

Short-Range Attack Missile (SRAM)

SRAM is a nuclear weapon to be fired from AMSA (B-1), FB-111, and B-52 bombers at ground defenses. Its dependence on bombers makes it a doubtful deterrent force. The elimination of these bombers would eliminate the need for SRAM.

Estimated annual savings: $156.30 million.

Estimated total savings: $1,500.00 million.

F-111 Long-range Fighter-bomber (formerly TFX)

The F-111 program will produce aircraft very similar to the existing F-105 and F-4, with the only major difference being range. However, present in-flight refueling techniques give the F-105 and F-4 long-range capabilities. The F-111 is inferior to existing aircraft in the air superiority role. It is not designed for limited-war missions and it has been indicated by the Army that the F-111 will not meet its close air support requirements.

The F-111 would only be effective in a limited nuclear or large conventional war with the Russians, since only the Russians have an air defense and interception system of sufficient grade to make the more advanced F-111 desirable. The F-111 therefore adds little value as a deterrent and would be of less value as a retaliatory weapon in the face of an all-out Soviet attack. The program should be abandoned.

Estimated annual savings: $563.30 million.

Estimated total savings: $7,000.00 million.[30]

Mark II Avionics Program

This program is concerned with the development of radars, computers and inertial equipment for use on F-111 aircraft. The elimination of the F-111 therefore would eliminate the need of the Mark II Avionics program.

Estimated total savings: $143.00 million.[31]

Shrike Air-to-Surface Missile

Both Shrike and the Standard ARM missile are to be used for anti-radar purposes. Sufficient ARM missiles are presently on hand. This duplication is considered to necessitate the dropping of the Shrike program.

Estimated annual savings: $21.60 million.[32]

Sparrow Missile

The Sparrow is an air-to-air missile to be used in aerial combat. In the event of a nuclear war, the extent of aerial combat will be almost zero, thus minimizing the need for such a missile. Therefore, the Sparrow is not considered to have any power as a deterrent and should be eliminated.

Estimated annual savings: $68.50 million.[33]

Hawk Missile

This air defense missile for the Army is having development problems. It is scheduled to be replaced by the SAM-D missile. Therefore, further development and procurement of the Hawk should be abandoned.

Estimated annual savings: $127.70 million.[34]

Lance Missile

This weapon is designed for the support of ground forces and would most probably replace older missiles at overseas U.S. bases. Unless we intend to start military operations in these areas, the Lance missile would not be needed.

Estimated annual savings: $83.40 million.[35]

Shillelagh Missile

This anti-tank missile would never see action in a nuclear war fought with ballistic missiles. Further production and development is considered to be irrational.

Estimated annual savings: $6.20 million.[36]

COSTS ASSIGNABLE TO THE INDO-CHINA WAR

(Note: All figures quoted from the United States Government Fiscal Year 1971 Budget and Appendix are 1969 actual expenditures unless otherwise noted).

Personnel

The total cost of pay allowances, clothing and subsistence for the approximately 3.5 million men constituting America's armed forces is $22,000 million. The average expenditure per man-year is, then, about $6,500.[37]

The average rate of compensation for civilian personnel may be found, in the same manner as above, to be $7,500 per man-year.[38]

On the basis of these figures, the approximate costs for pay and allowances to the 800,000 military combat and support personnel and the 250,000 civilian support forces associated with the Indo-China War can be calculated.

Estimated annual savings: $7,075.00 million.[39]

The Operations and Maintenance (OM) costs of the armed forces are also related to personnel outlays. The total OM expenditures less all civilian employee compensation (considered separately above) for all the services and the defense agencies is $15,807 million. About one-quarter of this amount may reasonably be attributed to operations in Southeast Asia (SEA).

Estimated annual savings: $3,671.00 million. [40]

Personnel total estimated annual savings: $10,746.00 million. (Note: Military personnel compensation, $5,200 million, plus OM costs, $3,671 million, yields a cost per man-year of $11,090. This compares favorably with C.L. Schultze's estimate of $12,000 per man-year[41] and the estimate of $13,000 per man-year published by Senator Kennedy.[42])

Ground, Air, and Naval Ordnance

The total expenditure in FY 1969 for ground, air and naval ordnance was $5,181.40 million. This figure is only slightly lower than that requested by Secretary of Defense Clark Clifford for FY 1970. According to Secretary Clifford's projected estimates of ordnance consumption in SEA, $4,800 million of the total requested should be designated for SEA operations.[43]

Estimated annual savings: $4,800.00 million.

Procurement and Other Equipment Costs

Procurement costs are the most difficult to accurately

evaluate. Numerous diverse obligations are codified under the heading of "Procurement" in the FY 1971 budget. In order to arrive at a reasonable estimate of the expenditures for new equipment specifically designated for SEA use, only those budget items pertaining to General Purpose Forces were included. All items related in any way to ordnance (see above) or ordnance support equipment were ignored. Further, any items providing for the purchase of missiles or missile support facilities and any expenditures thought to consist substantially of funds for strategic forces, were eliminated.

The total procurement expenditure, thought to be mostly for conventional forces usage, were determined to be $13,006 million.[44]

If represented as $3,776 per man-year, the fraction assignable to the 800,000 forces associated with the Indo-China War may be calculated.

Estimated annual savings: $3,021.00 million.
(Note: this estimate of $3,021 million compares favorably with estimates formulated by Charles L. Schultze, Former Director of the U.S. Bureau of the Budget. His estimates are as follows:

The replacement of 500 aircraft lost annualy at $3 million per aircraft, average.

Estimated annual savings: $1,500.00 million.

The replacement of Land Force equipment and supplies, U.S.

and ARVN, plus other procurement.

Estimated annual savings: $2,300.00 million.[45]

These figures include the replacement of 300 helicopters lost each year at $300,000 each, average.

Total estimated annual savings: $3,800.00 million.[46])

The total of above estimated annual savings realizable by the Department of Defense upon termination of SEA operations is:

Total estimated annual savings: $18,567.00 million.

Economic and Military Assistance to South Vietnam

Aid for Phase I of the modernization of RVN forces: $532.00 million.[47]

Aid for Phase II of the modernization of RVN forces: $120.00 million.[48]

Commercial Import Program aid to Vietnam: $320.00 million.[49]

Agency for International Development: 473.00 million.[50]

Combat Readiness, South Vietnamese Forces, Defense: $300.00 million.[51]

Estimated annual savings: $1,615.00 million.

Research and Development, Testing and Evaluation

Funds for the maintenance of a continuing Southeast Asia-oriented RDT&E effort.

Estimated annual savings: $525.00 million.[52]

Costs Not Included

Secretary of Defense Melvin Laird requested, in March 1969, an additional $77 million to support the currently high monthly rate of B-52 sorties in SEA. Despite their conventional usage in Indo-China, B-52 costs are included in Strategic Forces expenditures in the FY 1971 budget. Consequently, the amount in excess of $77 million dollars attributable to B-52 operations has not been determined and is not included.[53]

The costs of pacification and military construction would seem to be interlocking. Roads and railways which serve the population may be of a strategic use as well. The budgetary items covering military construction, in excess of $1,100 million, are not sufficiently well defined to allow a determination of the amount spent in SEA.

South Vietnam receives economic aid from the Food for Peace program of an undetermined amount.[54]

Some "mothball" ships have been converted for use in SEA. The costs associated with the restoration of these vessels have not been estimated.

There are more than 100 naval vessels stationed in SEA, including 3 aircraft carriers. The operating costs of these three carriers alone probably exceed $1,000 million. No estimates could be made of the total operating costs for all the vessels.

It was recently reported (The New York Times, June 8, 1970, p. 1) that the Department of Defense was paying the government of Thailand $50 million each year for the salaries and general support of the 11,000 Thai troops serving in Vietnam. At a similar rate, the aid to other allied troops (specifically Korean) in SEA would be approximately $250 million annually.[55]

Estimated annual savings: $300.00 million.

It is not possible to estimate the cost of operations in Laos, or of CIA and Air America operations in Thailand, Laos and Cambodia. It will be difficult to tell for some time the extent of military aid to Cambodian forces or its budgetary source.

Many of the weapons given to the South Vietnamese and Cambodian forces are of old design. They are probably taken from American forces elsewhere in the world. Replaced by newer weapons, this practice has modernized all our forces but has also incurred significant costs not ascribed to the Indo-China War.[56]

Although $18,567 million is a reasonable estimate of the costs assignable to the Indo-China War, it does not include a number of items. The most significant of these omissions would seem to be the B-52 operating costs, naval fleet costs, "hidden" U.S. and ARVN armament costs, and aid to allied forces.

COST FACTORS NOT INCLUDED IN ESTIMATES OF POSSIBLE REDUCTIONS, OWING TO DATA LIMITATIONS

The following categories of military forces were not examined for possible cost reductions traceable to revised military security criteria.

    Army   )   expenditures for operation and
)   maintenance other than those
    Air Force)   accounted for in the illustrations
)   above.
    Marine Corps)

Military Reserves and National Guard

Military Intelligence Operations ($2.9 billion in 1969)

The military base system

Cost of civilian dependents at overseas bases

Headquarters operations of the armed forces

Research and Development, Test and Evaluation

Selective Service System

Military Air Transport Service

Military Ship Transport Service

Military operations of the Central Intelligence Agency

Atomic Energy Commission military costs

Biological and Chemical Warfare

Foreign military assistance

Military construction

NOTES

[1] U.S., Congress, Joint Economic Committee, The Military Budget and National Economic Priorities, Pt. 1, Hearings, before Sub-committee on Economy in Government, 91st Congress, 1st Session, 1969 (Washington, D.C.: United States Government Printing Office, 1970), p. 358.

[2] Ibid., pp. 79, 80, 110.

[3] Ibid., pp. 110, 358.

[4] Ibid., p. 60.

[5] United States Naval Institute Proceeding, Vol. XCVI (May 1970), p. 391.

[6] Ibid., p. 390, and The Military Budget and National Economic Priorities, op. cit., p. 350.

[7] U.S. Naval Institute Proceedings, op. cit., p. 390.

[8] Ibid., p. 364.

[9] The Military Budget and National Economic Priorities, op. cit., pp. 111, 173, 352, and U.S., Congress, Senate, Joint Session of Armed Services and Appropriations Committees, Statement by Secretary of Defense Melvin Laird, fiscal year 1971, defense program and budget, Feb. 20, 1970 (Washington, D.C.: United States Government Printing Office, 1970), p. 149.

[10] U.S. Naval Institute Proceedings, op. cit., under Calendar of Events, July 1969, and The Military Budget and National Economic Priorities, op. cit., pp. 111, 173.

[11] U.S. Naval Institute Proceedings, op. cit., p. 461, and The Military Budget and National Economic Priorities, op. cit., p. 366.

[12] U.S. Naval Institute Proceedings, op. cit., pp. 391, 392.

[13] Ibid.

[14] Ibid., p. 393, and The Military Budget and National Economic Priorities, op. cit., pp. 273, 366, and U.S., Congress, Senate, Committee on Armed Services, Authorization for Military Procurement, Research and Development, Fiscal Year 1970, and Reserve Strength, Hearings, on S. 1192, S. 2407, and S. 2546, 91st Cong., 1st Sess. (Washington, D.C.: United States Government Printing Office, 1969), p. 57.

[15] U.S. Naval Institute Proceedings, op. cit., p. 393.

[16] Ibid., p. 391, and Authorization, op. cit., p. 55.

[17] National Journal (Center of Peace Research, Washington, D.C.), Feb. 7, 1970, p. 276.

[18] National Journal, loc. cit.

[19] Ibid.

[20] The Military Budget and National Economic Priorities, op. cit., p. 110.

[21] Statement of Secretary of Defense Melvin R. Laird, op. cit., p. 138.

[22] Ibid., p. 276.

[23] Ibid., p. 139.

[24] U.S., Congress, Senate, Joint Session of Armed Services and Appropriations Committees, Statement by Secretary of Defense Melvin Laird, fiscal year 1971, defense program and budget, Feb. 20, 1970 (Washington, D.C.: United States Government Printing Office, 1970), p. 113.

[25] U.S., Congress, Joint Economic Committee, The Military Budget and National Economic Priorities, Hearings, before subcommittee on Economy in Government, 91st Cong., 1st Sess., 1969 (Washington, D.C.: United States Government Printing Office, 1970), testimony of Charles Schultze and Robert Bensen; and National Journal (Center of Peace Research, Washington, D.C.), Jan. 3, 1970; and National Journal, Feb. 7, 1970; and statement by Dr. Leonard Rodberg (University of Maryland), March 25, 1970; and The New York Times, May 18, 1969, p. 72.

[26] National Journal, Jan. 3, 1970, p. 37.

[27] Cameron, Juan, "The Case for Cutting Defense Spending," Fortune Magazine, August 1, 1969.

[28] National Journal, Feb. 7, 1970, p. 38; and U.S. Department of Defense, Report to the Congress by the Comptroller General of the United States - Status of the Acquisition of Selected Major Weapons Systems, Department of Defense B-163058, Feb. 6, 1970 (Washington, D.C.: United States Government Printing Office, 1970), p. 36.

[29] National Journal, Jan. 3, 1970, p. 38; and National Journal, Feb. 7, 1970, p. 276; and Report to Congress by the

Comptroller General, op. cit., p. 36.

[30] National Journal, Feb. 7, 1970, p. 276; and Melman, Seymour, "Memorandum on the TFX," Jan. 1964.

[31] U.S., Congress, Joint Economic Committee, The Economics of Military Procurement, Report of the Subcommittee on Economy in Government, 90th Cong., 2nd Sess., 1969 (Washington, D.C.: United States Government Printing Office, 1969), p. 23.

[32] National Journal, Feb. 7, 1970, p. 276; and National Journal, Jan. 3, 1970, p. 44.

[33] The Military Budget and National Economic Priorities, op. cit., p. 593; and National Journal, Feb. 7, 1970, p. 276.

[34] National Journal, Jan. 3, 1970, 44; and National Journal, Feb. 7, 1970, p. 276.

[35] The Military Budget and National Economic Priorities, op. cit., p. 593; and National Journal, Jan. 3, 1970, p. 37; and National Journal, Feb. 7, 1970, p. 276.

[36] The Military Budget and National Economic Priorities, op. cit., p. 593; and National Journal, Feb. 7, 1970, p. 276.

[37] The Budget of the United States Government, Fiscal Year 1971, Appendix (Washington, D.C.: United States Government Printing Office, 1970), pp. 265-68, Item 10, and FY 1970 pay increases.

[38] Ibid., pp. 274-279.

[39] Schultze, Charles L. (with Hamilton, Edward K., and Schick, Allen), Setting National Priorities, The 1971 Budget (Washington, D.C.: The Brookings Institution, 1970), p. 49, and attached projections of pre-Vietnam manpower trends (see pages 31 and 32).

[40] The Budget ... 1971, Appendix, op. cit., pp. 274-279, Items 21.0-26.0, 31.0, 32.0, 41.0, 42.0.

[41] Schultze, Charles L., op. cit., p. 49.

[42] U.S., Congress, Senate, Committee on Labor and Public Welfare, Postwar Economic Conversion, Hearings, 91st Cong., 2nd Sess., 1970 (Washington, D.C.: United States Government Printing Office), p. 425.

[43] The Budget ... 1971, Appendix, op. cit., pp. 283-92; and Statement of Secretary of Defense Clark M. Clifford, The Fiscal

Year 1970-74 Defense Program and 1970 Defense Budget (Washington, D.C.: United States Government Printing Office, 1969), pp. 72-74.

[44]The Budget ... 1971, Appendix, op. cit., pp. 273-293, Budget Program by Activities Items: Army, p. 283, items 1, 2, 5-8; Navy, p. 285, items 1-3, 5-7, and p. 287, items 1, 2, 5-7; Marines, p. 289, items 2, 4-6; Air Force, p. 290, item 10, and p. 292, items 2-4; Defense Agencies, p. 293.

[45]Schultze, Charles L., loc. cit.

[46]U.S., Congress, Senate, Committee on Armed Services, Authorization for Military Procurement, Research and Development, Fiscal Year 1970, and Reserve Strength, Hearings, 91st Cong., 1st Sess., on S. 1192, S. 2407, and S. 2546 (Washington, D.C.: United States Government Printing Office, 1969), testimony of General McNickle, p. 1631.

[47]U.S., Congress, Senate, Armed Services Committee, Defense Report, Statement by the Honorable Melvin R. Laird, The Secretary of Defense, 91st Cong., 1st Sess, March 19, 1969 (Washington, D.C.: United States Government Printing Office, 1969), p. 17.

[48]Ibid., p. 19.

[49]The Budget of the United States Government, Fiscal Year 1971, U.S.G.P.O., 1970, p. 94, FY 1971 estimate.

[50]The Budget ... 1971, Appendix, op. cit., p. 80.

[51]Ibid., p. 300, FY 1971 estimate.

[52]U.S., Congress, Senate, Joint Session of Armed Services and Appropriations Committee, Statement of Secretary of Defense Melvin R. Laird, Fiscal Year 1971, defense program and budget (Washington, D.C.: United States Government Printing Office, 1970), p. 72, current FY 1970 level.

[53]Ibid., p. 35, and The Budget ... 1971, Appendix, op. cit., p. 273.

[54]The Budget ... 1971, op. cit., p. 97.

[55]Statement of Secretary of Defense Clark M. Clifford, op. cit., p. 74.

[56]U.S., Congress, Senate, Armed Services Committee, Statement of Secretary of Defense Robert S. McNamara, Fiscal Year 1969-73 Defense Budget and 1969 Defense Budget (Washington, D.C.: United States Government Printing Office, 1968), pp. 88-94.

A COUNTERCOMBATANT DETERRENT?
FEASIBILITY, MORALITY, AND ARMS CONTROL

Bruce M. Russett

AN ALTERNATIVE TO CITY-BUSTING?

Since the beginning of the cold war, the keystone of American strategic planning has been the principle that the ultimate deterrent to a Russian attack had to be the certainty of American retaliation against Russian cities. The Russians seemingly have adopted the same position vis-a-vis the United States, usually to the point of denying vigorously any possibility of limiting central war to counterforce strikes alone. A typical relatively recent formulation was expressed by former Defense Secretary McNamara in his 1968 "posture statement" to Congress:

> (It is) the clear and present ability to destroy the attacker as a viable 20th Century nation and an unwavering will to use those (Assured Destruction) forces in retaliation to a nuclear attack upon ourselves or our allies that provides the deterrent, and not the ability to partially limit damage to ourselves. [1]

Several points about this statement must be noted. One is the implicit precept of retaliation -- the eye for an eye principle found in international law as well as in private behavior. Even so, this particular application, that one will retaliate against <u>civilians</u> for the acts of a <u>government</u> which we widely assume does not accurately reflect their wishes (to be sure, perhaps for acts committed against one's own civilians) might give us pause. Second, the deterrent umbrella is extended also to cover America's allies. This is quite in accord with generally accepted ideas of the right of the strong to protect the weak, and the international principle of collective defense. Finally, there is the clear implication, never completely disowned by American leaders, that even a hypothetical counterforce strike initiated by the enemy might be met with a countercity attack in "retaliation." The problem is one of retaining an "assured destruction" capability; if the enemy were too successful in his counterforce attack, what residual American missiles were left might be turned against Russian cities even though American cities had so far been left unscathed. This threat, of if "necessary" doing very great damage in response to <u>any</u> kind of large nuclear attack on the United States, is at the heart of the "assured destruction" principle. Despite all the doctrine and preparations for controlled tit-for-tat strikes against the enemy, the threat of initiating countercity warfare, under dire circumstances, remains in the background.

In this paper I want to challenge the idea that deterrence must rest on an assured capability of destroying an enemy's cities and non-combatant populace. I want to do so on the grounds that what I shall call a countercombatant strategy is militarily feasible, and also is preferable on widely accepted moral grounds. In short, I want to present a case that Americans could adopt a new deterrent policy that would ease the moral qualms many of them have about current strategy, without leaving them less secure than at present. I candidly admit that my search for an alternative deterrent strategy is rooted in a moral revulsion against plans deliberately to kill large numbers of civilians in case of central war, but I would not impose that view on others, nor would I trouble to pursue the question in public at all if I were not coming to the conclusion that there _is_ a viable alternative short of pacifism.

I am somewhat diffident about making my proposal. It is not based on any new computations or empirical evidence, merely on what I think is a different perspective on familiar matters, possibly made more appropriate by some developments in weapons acquisition in recent years. By many tests it is not a radically new proposal. For example, it does not in itself challenge the widely -- but increasingly less universally -- accepted belief than in contemporary superpower politics <u>some</u> capability to inflict vast damage on the other remains a necessary ingredient of a war-deterring posture. It is not addressed to, nor does it anticipate, a great reduction in the number of nuclear delivery vehicles the United States must maintain. Nor will it in itself lead to major arms control agreements about numbers and types of weapons. It would be consistent with many such agreements, but not with others. While these things are important, they should not always be dominant. In bypassing them my proposal is perhaps a conservative one. But what it does do is require us to think less about deterrence as simply an ability to inflict casualties, and more about what kind of casualties might achieve politically significant results. Such a rethinking might have radical and highly desirable consequences.[2]

RESTRAINTS AND THEIR EROSION

It is not easy to draw a clear, unambiguous line between combatants and non-combatants. Nor have warriors always tried very hard to make the distinction; no precise codification of rules was ever fully accepted. But in the years following the enormous civilian suffering of the Thirty Years War various limits were generally respected. Wars were fought for restricted objectives, and a combination of self-interest, humanitarianism, technological incapacity, and the inability of aristocratic regimes to motivate their soldiery for long or intensive wars provided important restraints.

Some of these limits were eroded during World War I, with a partial failure of the international law of war, especially at sea in Germany's submarine warfare and Britain's embargo. International law on the matter of bombing non-combatants was vague; the strictures on the conduct of war evolved long ago and even now apply primarily to land warfare.[3] Whatever legal restraints might have been appropriate to aerial warfare were substantially loosened. In this first major conflict fought with aircraft and dirigibles, aerial warfare was conducted against civilians as well as against military targets. Subsequently, military theorists of various nations, such as Giulio Douhet, Billy Mitchell, and Air Marshal Sir Hugh Trenchard all advocated the use of air forces against civilian populations. Their doctrine, plus the World War I precedents and then such actions of the 1930s as those of the Italians in Ethiopia, the Spanish Civil War experience, and Japanese bombing of Chinese civilians prepared the governments of Europe for the worst in 1939. Although military people especially in England objected to

Trenchard's plans for population attacks, it came to be considered inevitable that such methods would be turned on Britain and France once the war actually was begun. Elaborate civil defense preparations were made, and an exaggerated fear of air attacks on their cities seems to have been a significant factor in those governments' unwillingness to stand up to Hitler at Munich.

But they were pleasantly surprised by what actually did and did not happen in the first part of the war, known because of the restraints as the period of "phoney" war. For almost a year Hitler carefully avoided making attacks on civilian areas in France and England, limiting the Luftwaffe to strictly military targets. He was less restrained in the Blitzkreig on Poland, but even there civilian deaths occurred largely as a result of tactical air support of ground forces. Since Hitler was no great humanitarian, his motives were governed by self-interest -- he feared the effect that allied bombing of Berlin and other German cities might have on German morale, and so wanted to avoid initiating an exchange of city strikes. Furthermore, he hoped for an early compromise peace, and by withholding his air arm could both avoid antagonizing his enemies unnecessarily and also retain an implicit threat to initiate something worse if they did not cease fighting. Britain and France, for their part, recognized their inferiority in aircraft and knew that if anyone did start attacking cities they themselves would suffer more damage than they could return to the Germans.

These limits deteriorated somewhat during the first half of 1940, notably with the German bombing of Rotterdam in May. The ultimate collapse of restraints during and after the Battle of Britain and Blitz is well-described by George Quester; apparently the breakdown ought not to be blamed entirely on the Germans, as is commonly believed.[4]

In any case, during the rest of the war saturation terror attacks became routine. On the Axis side they included the V-1 and V-2 bombs late in the war. As for the Allies, the British habitually attacked at night when precision bombing was impossible, deliberately directing many of their strikes against residential areas so as to undermine popular morale. By comparison, air raids by the United States were largely conducted during daylight hours and directed to industrial targets, which blows were thought to damage the German war effort more than terror attacks could. But the Americans too, as in the attack on Dresden, occasionally went after civilians deliberately. And if Americans can claim a few credits for restraint in the air war over Europe, the firebomb raids on Japanese cities (in which a ring of fire was carefully built to trap people inside), not to mention Hiroshima and Nagasaki, remove much virtue from the account.

The point is not that most of World War II was fought with little restraint on bombing civilian areas -- that is obvious. Rather, it is equally important that for quite some time each side did limit its actions and was aware that the other side was doing the same. Neither moral nor self-interest restraints were entirely put aside at the outset. The former were still sufficiently in many peoples' minds that President Roosevelt could characterize the earliest, mildest German air attacks in Poland as "inhuman barbarism which has profoundly shocked the conscience of humanity."[5] The limits as observed, for whatever reasons, should be remembered as well as the unrestricted bombing of the later years.

After World War II the original restraints were forgotten. On the basis of their own precedent in Japan, American military planners simply assumed that in future wars nuclear weapons would be used against cities to destroy the enemy's economy, society, and popular morale. In addition, the threat of such usage was to constitute the primary deterrent to war. In the 1948-49 controversy over the B-36 a number of United States Navy officers argued that countercity war was immoral, but their objections were largely dismissed as rationalizations to further their Service's interests. The basic countercity strategy then remained essentially unquestioned until the late 1950s, in other countries as well as in the United States. Since World War II every major power, either in that conflict or in colonial wars thereafter (for example, the French in Algeria and against Tunisia), made deliberate air attacks on civilians.

Some very important and promising steps toward redefining limits on aerial warfare were nevertheless taken in the past decade or so. They began with the idea of tactical nuclear warfare. While the dream of confining nuclear weapons to tactical uses has by now been pretty thoroughly abandoned, it helped in defining the idea of limited strategic or central war. To some strategists it became not just something to be pursued if the opportunity to avoid general destruction should present itself, but even a feasible policy. In 1962 Secretary McNamara reached the point of making his now famous Ann Arbor statement: "...Principal military objectives, in the event of a nuclear war stemming from a major attack on the Alliance, should be the destruction of the enemy's military forces, not of his civilian population."[6] Thus massive retaliation was to give way, even in response to "a major attack on the Alliance," to controlled graduated response and the hope of reciprocation from the enemy.

While the distinction between counterforce and countercity targeting remains important in American strategic doctrine, McNamara's attempted annunciation of a "no-cities" strategy as a viable retaliatory policy has largely been rebuffed and withdrawn. His own 1968 statement witnesses an apparent return to "assured destruction" of civilian population centers as the perceived sine qua non for a credible deterrent. McNamara's Ann Arbor policy was widely attacked for many reasons. Primary among them, however, was the fear that the Russians would interpret it, and the weapons procurement to implement it, as indicating an American attempt to secure a high level of "damage limitation" -- an emphasis on counterforce so strong as to seem to permit the United States to engage in a pre-emptive or even a preventive nuclear strike against the Soviet Union. That could be profoundly destabilizing, and might well bring on the horrible war McNamara hoped to avoid. It became widely assumed that if a counterforce capability was good enough to be "worth" having for damage limitation, it would bring too great a temptation, or too great an apparent temptation as seen by the opponent, to use it in a first-strike capacity. McNamara's argument for counterforce retaliation was seen as not capable of being distinguished, in practice, from the ability to initiate war at what might, in extremis, be considered a politically acceptable cost. If not that, a damage-limiting policy might at least lead to an accelerated arms race with no ultimate gain; the Russians would take the necessary steps to counteract it.

The strategy embodied in McNamara's Ann Arbor speech was supported by many strategic analysts, perhaps most notably by Thomas Schelling in a model of clear thinking. Schelling noted that nuclear war was not a shooting gallery where one had to shoot at something. A doctrine of <u>not</u> attacking cities did not <u>necessarily</u> mean that the opponent's nuclear delivery vehicles therefore had to be attacked en masse. On the contrary, one might choose substantially, at least in the early stages of a war, <u>not to fire at anything.</u> The opponent's cities could be left for a while as hostages to his good behavior; so long as he avoided deliberately striking our civilians we would reciprocate. "The basic reason for not destroying the cities is to keep them at our mercy."[7] Schelling then went on to show why a city-avoidance strategy was appropriate for adoption not only by a power with nuclear superiority (then the United States) but at least equally by a power with only equivalent or even inferior nuclear strength. He demonstrated that its desirability did not depend on being able actually to <u>disarm</u> the opponent, since the intention was to bargain away his willingness to use capabilities he might still retain. Nor, he argued, did it depend on the Russians following the substantial American lead of locating missile sites outside the boundaries of populous cities. While the latter might be a desirable condition from the Russian as well as the American point of view, once the shooting gallery notion was disposed of such location was not essential to an American pursuit of city-avoidance.

Schelling's argument was subtle and not widely accepted. At the time of McNamara's speech much of the Russian nuclear retaliatory force was in fact vulnerable to an American first strike, and discussion of a city-avoidance strategy, especially one so subtle, was feared to be unsettling.[8] Most important perhaps was the apparent lack of reciprocation by the Russians. Schelling's strategy, and McNamara's, did depend for its acceptability on Russian willingness also to consider avoiding cities, and to be capable of receiving the tacit or explicit bargaining messages in war. Virtually all public statements by Soviet military analysts dismissed the possibility of a city-avoiding nuclear war. A more relevant answer might have been contained in Soviet weapons-procurement policy, had they clearly moved toward the acquisition of missiles of high accuracy and low yield. But in fact the procurement messages were at best ambiguous. One does not hear much about it today.

HOW MUCH IS ENOUGH?

Nevertheless, I intend here in part to revive the notion of counterforce strategy as plausible, in the form of what I call a countercombatant strategy. The strategy does not imply a first-strike capability or anything resembling it; indeed it assumes that, given prudent vigilance on both sides, such a capability is not a realistic possibility for either super-power. Nor does it imply an ability to "win" a nuclear war, whatever that means -- on the contrary it assumes that even in "losing" a countercombatant war the United States still could inflict such a level of damage on the Russians as to make the war's initiation unprofitable. Nor finally -- and here it differs markedly from the Schelling-McNamara strategy -- does it <u>depend</u> for success upon using the enemy's cities as hostages, on bargaining, or on reciprocation. While bargaining might well be attempted during its execution, it is nevertheless conceivable that an American government <u>might</u> want to hold to a countercombatant policy even if the enemy did not.

Again, the standard assumption of contemporary strategic theory has been that deterrence rests upon a capability of wreaking "assured destruction" which would "destroy the attacker as a viable 20th Century nation." One sees different figures, but under Secretary McNamara the operational measure of such destruction seems to have been on the order of 20-25 per cent of the Russian population and half of Russian industry.[9] Anything less was thought to constitute a temptation to aggressive or reckless Soviet leaders.

It is not entirely clear just what was the basis of these estimates. In part, doubtless they represent an estimate of a level of destruction that would make it impossible to reconstitute anything like the present social and political system in the Soviet Union, a level which would move beyond just very great loss to something approaching chaos among the survivors, considerable number though they might be. It surely represents an understanding that, as Herman Kahn put it, the Russians lost about 20 million people in World War II (10 per cent of the population) and doubtless considered it a good bargain, hence the level of assured destruction must be significantly higher.[10] For technical reasons it probably was approximately the level of damage beyond which it became increasingly costly to go. Because of diminishing marginal returns and the ability to achieve it with relatively few surviving weapons, the difference between Russian casualties from a first-strike and a second-strike was minimized -- a crucial deterrent criterion. In any case, it assumes that the surest means of deterring Russian leaders is through a threat to their populace and industry. Given some of the folklore that all Bolsheviks are as callous toward their people as Stalin seemed to be this may seem a bit curious, though perhaps less so if we consider that a good Bolshevik presumably would be properly solicitous about his power base -- populace and industry.

The reasoning is in fact highly questionable. Despite all the computations of megatonnage, defensive weapons, and population densities, it is composed of as much mythology as science. We simply cannot <u>know</u> what destructive capability is required to constitute effective deterrence. It may be that a very large assured destruction capability is necessary, or it may be that "merely" the collateral damage to civilian, industrial, and other military targets incurred from a counterforce strike would suffice. Furthermore, concern for his more narrowly military power base might weigh as heavily with a "rational" Russian leader who was contemplating an attack on the West. Severe damage to the armed forces of the Soviet Union, which are strongly relied upon for the maintenance of internal security as well as bloc cohesion, might be as unacceptable as the loss of cities.

Moreover, the level of retaliatory damage that a nation contemplating a first strike will accept is a <u>variable, not a constant.</u> The prospect of incurring great damage will be acceptable to a government if all other courses of action seem to lead to extremely undesirable outcomes. On the other hand, if "peace" offers half-way decent prospects the government will not follow a course of action promising anywhere near the level of probable war damage that it would accept if its back were against the wall. Thus in considering whether our nation has "enough" deterrent capability, we must always consider the policy context -- what options are presented to the government one wants to deter. For example, imagine that in the Cuban missile crisis President Kennedy, instead of leaving Premier Khrushchev a relatively graceful exit, had tried to humiliate him and publicly to expose Soviet weakness in a way that would greatly have diminished Soviet influence in the world. Under conditions such as these, continued "peace" would have looked less desirable, and at least relatively speaking nuclear war would have become more "thinkable." A level of deterrent capability that had previously been more than enough might no longer have sufficed, and the Russians might have become willing to launch a first-strike attack on the United States even in the knowledge that they would suffer severe retaliation.

I blush to make once again so obvious a point, one that was long ago stated clearly by such strategic analysts as Kahn, Schelling, and Albert Wohlstetter and expressed formally by such a variety of observers as Daniel Ellsberg, Morton Kaplan, J. David Singer, Glenn Snyder, and myself. But for many would-be strategists it does seem necessary. Deterrence results from making peace better as well as from making war look worse. And super-power politics is not a zero-sum game. Thus it is prudent for a would-be deterring power not to present its opponent with the prospect of very undesirable outcomes in return for forbearance from taking the first-strike option. It becomes extremely dangerous to seek clear-cut "victory" on crucial issues <u>even</u> by peaceful means, for fear of driving the other side to initiate war.

We cannot know, without getting inside the Soviet leaders' heads, what they will accept under various conditions. Certainly the Western academic specialists on the Soviet foreign policy do not agree as to their degree of aggressiveness or on what kind and degree of damage is necessary to deter them. But under "normal" circumstances of peaceful coexistence it would seem that the standard rule of thumb, 25 per cent of the Russian population, is a higher-than-necessary level of assured destruction most of the time. Only under conditions of <u>very</u> grave threat to the country or to communist rule would the Russian leaders be likely to launch an attack at that price. Kahn's example of the acceptable cost of resistance to German invasion is compelling precisely because the alternative seemed so ghastly -- especially when the Soviet people got a good look at Nazi behavior. Providing that the United States pursues a policy of reasonable restraint it might be secure with a somewhat less awful threat capability.

MILITARY TARGETS ONLY: FEASIBILITY

I will now propose that the United States could, without increasing the likelihood of war, significantly reduce, though hardly eliminate, the severity of possible war by adopting a policy of using its nuclear weapons only in what I call a countercombatant capacity. By countercombatant I mean to include first all of the basic nuclear striking forces of the enemy and their immediate support facilities: missile silos, military airbases, nuclear submarines and submarine bases, air-defense and ABM systems, and weapons-oriented atomic energy plants. More than than, it would include internal security forces, all military bases of any kind, and those transport facilities devoted primarily to the movement of troops and military supplies. Since the Russian leaders will be concerned about maintaining domestic order, and with regulating the behavior both of the states on their Eastern borders and of their Chinese neighbors, these "tactical" forces <u>will</u> matter in the context of a nuclear war that lasts more than a very few days and in the context of picking up the pieces later. So too, if not during the nuclear war itself, weapons and military equipment-manufacturing plants would matter, and I include them as appropriate targets for a retaliatory strike. It is for this reason that I use the term countercombatant rather than the more familiar counterforce -- it implies something more extensive than retaliation against nuclear striking forces, but less than a countercity strike.[11] (Even so the term is imprecise since many or most workers in weapons plants will not be combatants in the usual sense of the term, but so far I have been able to think of nothing better. They are at least adults who know what they are making and do so with at least some minimal degree of consent.)

Moreover, it might well be stated (and intended) American policy to attack these targets almost <u>wherever</u> they may be found, even near or within big cities. The United States would presumably develop a capability that gave it very high confidence of destroying those targets. But it would take great care to use <u>only</u> the amount of force necessary to destroy those targets. "Clean" weapons, of low yield and high accuracy, could be designed for the purpose of honestly striving to reduce destruction of population centers to whatever level is unavoidable in destroying what I have designated as military targets. "Bonus effects" would <u>not</u> be sought.

The extent of collateral damage incurred in a given instance depends upon the choice of targets, the megatonnage of the warheads, the amount of radiation released due to the composition and explosion height of the warheads, and the accuracy of the missiles. Thus collateral damage is considerably diminished if military installations in major population centers are avoided and small "clean" air-burst nuclear weapons with high accuracies are used. To the extent that installations within population centers are hit, the need for small, clean, accurate weapons is magnified if collateral damage is to be minimized.

The accuracy of intercontinental missiles has improved tremendously in recent years. The details change constantly and are in any case classified. But if we assume missiles with a CEP of about a quarter of a mile, quite low bomb yield, with comparatively low heat and blast damage to all but the surrounding areas, would suffice to knock out such targets providing the targets were not greatly hardened.[12] Collateral damage can, therefore, be limited, except for the many persons who will probably die from long-term radioactive fallout almost regardless of the location of the blast. Providing that force targets remain vulnerable and strategic emphasis is upon the development of low-yield weapons with high accuracies, for damage to civilians the more significant element in target selection is the type of target rather than target location. Thus it seems plausible that a countercombatant strategy could be implemented, indeed even to the extent of striking at a wide range of military facilities without much regard to their location, without inflicting nearly such a high level of civilian casualties as would be implied by a deliberate countercity strategy. But at the same time we must not deny that many civilian deaths would occur despite an American effort to keep them down.

Thus in the event of a Soviet attack on the United States or its allies, an event calling forth American nuclear retaliation, the following situation could be expected to emerge in the Soviet Union: The great majority of Russia's population, even the urban population, even of Russia's non-military industrial capacity, would survive. But its war-fighting capacity, and its capability for using force to maintain internal order, most certainly would not. Its tactical military capability, as well as its strategic forces, would be substantially gone. The Soviet Union's ability to defend itself from its neighbors, even the small and now much weaker states of Eastern Europe, would be destroyed. To make this particularly painful the United States might strike, with particular loving care, Russian bases and armed forces along the Chinese border. In effect, the penalty for a Soviet attack on the West would be Soviet impotence vis-a-vis their Asian neighbor. It is not at all obvious to me that this would be a less effective deterrent against

Russian adventurism than is the current threat to destroy many of the Russian people and the majority of their industry.[13]

Note once again, the proposed countercombatant strategy in _no_ way implies developing a force capable of delivering a first strike, and is not therefore destabilizing to the nuclear balance. It is indeed unimportant that its damage-limiting capacity be great. The intent is not to reduce Soviet capacity of harming the United States, but to reduce to an unacceptable level Soviet capacity to defend themselves tactically and/or strategically. <u>For these purposes</u> it does not matter whether the Russian missiles are destroyed on the ground or whether they really are used up against the United States. Even if substantial Soviet nuclear forces remained, they would be useless against internal dissent, and of only marginal utility in holding Eastern Europe, where ground troops would be needed for occupation. In a world where China or other third powers had significant nuclear and conventional capabilities, the political potential of nuclear weapons alone would be sharply circumscribed. To destroy the effectiveness of Soviet conventional forces we do not have to kill all their troops. It would be enough to hit bases, supply centers, military administration centers, and marshalling centers -- in short, to destroy their ability to <u>use</u> troops.

Finally, the strategy does not require symmetrical damage-inflicting capability between the two super-powers, nor would it necessarily allow the United States to "win" the war. Nuclear war is not a football game where the side which runs up the highest point total emerges with a victory worth having. It is patently absurd to insist that to be deterred the Russians must have more damage done to them than they are able to do to others; whatever the relative body count they would not in any meaningful sense win the exchange I have described.

Does a countercombatant strategy require Russian reciprocation? Insofar as protecting the lives of American civilians should a nuclear war occur is a major American goal -- and I am sure it is -- then clearly Russian reciprocation is highly desirable. But as a strategy for <u>deterring the outbreak</u> of such a war reciprocation is <u>not</u> necessary. It assumes, and I think plausibly, only that the Russian leaders would gain too little satisfaction from <u>anything</u> they might do by initiating war with the United States to compensate them for the military losses they could expect to suffer. On the other hand, in the <u>actual conduct of a war</u> it surely would be hard to maintain a purely countercombatant American strategy in the face of obvious Russian non-reciprocation. "Irrational" forces, including the demand for revenge, might well expand the retaliation into a countercity one. Furthermore, the advantages of making American restraint conditional upon Russian war-fighting conduct are obvious. It could provide a major, and perhaps crucial, bargaining counter to insure Russian restraint. I think I personally would prefer the complete abandonment of _any_ American intention to strike civilians deliberately, on the ground that it is unjust to punish civilians (including children) for acts of their government. But many people would surely retort that it is immoral to throw away a bargaining counter that might protect our own civilians.

Many varieties of reciprocation or non-reciprocation are possible. As a deterrent strategy the Russians might accept it and work both actively and visibly to implement it, they might be much more cautious or ambiguous, or they might reject it and refuse to change their policies. There is some reason to hope they might accept it. The Russian leaders have repeatedly declared they are in a struggle with the American government and capitalism, not the American people. They have consistently pictured ABM systems as desirable for preserving human lives. Despite their failure to respond favorably to McNamara's no-cities initiative, their statements about targeting focus on "the economic base of war, government and military administration, and groupings of enemy troops" -- but not countercity per se.[14]

While their definition of war-related industry probably would be more inclusive than the one I am suggesting, that still is not quite the same as pure counterpopulation strikes. I would rather see them hit the River Rouge Ford plant than hit Detroit.

Nevertheless, they might reject the American initiative, with an extreme form of rejection taking the form of an attempt to exploit it. After all, they _could_ choose to put a large proportion of their military forces of all types in greatly hardened sites directly within cities. In so doing they might well force the United States government to consider those sites to be primary targets despite the perhaps tens of millions of civilian deaths that would follow their destruction. (A key element of a countercombatant strategy, as it was not for Schelling-McNamara no-cities, is greatly reducing the Russians' ability to deter the Chinese and control their own citizens.) Or they might, through a combination of dispersion and provision for high mobility, vastly increase the number of military targets we would have to hit. But the moral burden would then be clearly shifted to the Soviet leaders; if we then abandoned the effort to spare civilians we would at least know that we had tried. Frankly I find it hard to imagine why the Russians would think they could exploit the American countercombatant initiative successfully. The effort would lead back to an American policy of retaliation against weapons _and_ cities, not to an abandonment of our reliance on the nuclear deterrent.

SOME MORAL CONSIDERATIONS

Men reason from a variety of moral premises, and for that matter simply try to avoid difficult moral choices whenever possible. To many it seems that the American government has no feasible alternative to its present policy. Few lines of thought are more painful than facing an evaluation that one's acts are immoral if there in fact seems nothing else one would do. But what I have argued above is that there may indeed be an alternative; hence even for one who on moral grounds rejected the idea of countercity deterrence it could become possible to have one's cake and eat it too. If a policy of countercity retaliation were truly not essential to national preservation, then a failure to discard that policy would be much more widely regarded as immoral. Let us then briefly review some standard judgements as to what constitute morally acceptable acts of warfare.

A completely pacifist position may result from a philosophical and moral predilection for non-violence, a rejection in principle of force as an instrument of national policy, a belief in the spiritually regenerative effect of a non-violent response to violence, or an overriding concern for the preservation of human life. Or, a position which is pacifist for all practical purposes may emerge from considerations of the principle of proportionality, a presumption that in a modern thermonuclear war the costs must inevitably outweigh the gains. At the other end of the spectrum is the view which says essentially that war, and any act in war, is justifiable if it seems to serve the "national interest," or that rightness depends solely upon the ends being sought rather than on the methods used to obtain them. Between these, a variety of positions still are available.

For those who accept the use of force as a legitimate instrument of state policy in many but not all circumstances, there are two principal moral foci for viewing its limitations. One is concerned with the norms that govern <u>recourse</u> to war, and the second adds to them some norms to govern the <u>conduct</u> of war. [15] The first, which concentrates on what conditions justify an initial resort to physical violence, is typically less concerned with the manner in which the conflict is conducted, once begun. In the American philosophical tradition a "just war" often is considered only to be one undertaken in self-defense. [16] Self-defense by this definition includes defense of one's allies in keeping with a formal commitment, assistance to a small power under the principle of collective security when authorized by an international organization like the United Nations even if there is no treaty commitment, or assistance to another government in response to its request for aid. Furthermore, the self to be defended is generally construed broadly to include not merely the physical territory but also the values and way of life which are believed to characterize the state. No grievances, however severe, would by this definition in themselves justify the initiation of war; grievances should always be subject only to negotiation or arbitration, or ultimately lived with in the hope they will become more tolerable through the evolution of circumstances.

On the other hand, once a war in self-defense is undertaken, limits on both the political objectives to be achieved and the means to be used in pursuit of them become hard to establish. The erosion of such limits helps destroy the opponent's incentive to make peace. The view emphasizing the justice of our recourse to war may have played a role in the breakdown of restraints on American conduct of air war in World War II, though as I noted above the breakdown was universal and by no means solely or initially the responsibility of this nation. I do suspect, however, that this view combined with the World War II precedents to make it easy for the doctrine of countercity deterrence to be accepted after the war, during the period of American nuclear monopoly.

We should also note the most common communist view, according to which a war need not be undertaken in self-defense to be justifiable, but may be perfectly right if its purpose is to redress certain grievances, namely class oppression or national subjugation. In this respect it differs widely from the classical American doctrine. Even for the communists, however, to be just the war must not have a reactionary effect. Specifically, a nuclear war which would result in the annihilation of capitalist <u>and</u> socialist civilizations would not be initiated. Not just any hypothetical war undertaken by a communist or third-world country would be permissible. Differing interpretations of the likely result of nuclear war underlay some aspects of the public differences between Russian and Chinese leaders during the past decade.

The second and quite different position stems from the tradition of Christian moralists.[17] It takes off from a recognition that attempts only to limit the <u>resort</u> to war are subject to abuse, and seeks to supplement them with rules for the <u>conduct</u> of war. Relevant elements of this tradition regarding the conditions under which a war may be "just" include:

1. Those who resort to war must have a "right intention;" this means substantially that they must do so in self defense or to correct a legitimate grievance (the definition of which varies).

2. The injury the war is intended to prevent must be real and "certain." (Obviously this last is elastic.)

3. War must be undertaken only as a last resort.

4. The measures employed in the war must themselves be moral. (Prisoners must be fairly treated; the inviolability of non-combatants respected, etc.)

5. The seriousness of the injury to be prevented must be proportional to the damages that are inflicted.

6. There must be reasonable hope of success; i.e., hopeless resistance cannot be justified.

The principle of double effect is frequently applied to the evaluation of particular courses of action during the war. That is, the evil done by any act must not be <u>willed</u>, but only tolerated. This criterion obviously is subject to casuistic abuse, but it does still offer the potential for important restraints. By precisely this principle it would be permissible to bomb missile sites near cities, even though some civilians would be killed by the bombing, provided that the above conditions were met (especially the principle of proportionality, point number 5). But saturation bombing, deliberate aerial bombardment of residential areas, or use of larger bombs than were strictly necessary to destroy the military target would be condemned.

So would the deliberate initiation of attack on predominantly civilian targets. In a further elaboration of the "double effect" argument, any good achieved must follow directly from the <u>act</u>, not from the evil effect. This may really sound convoluted, but it is clear enough, for instance, regarding use of the atomic bomb against Hiroshima and Nagasaki. Though it may well have shortened the war and avoided hundreds of thousands of civilian casualties in savage ground combat, that would not be enough. The war presumably ended as a direct result of the intended civilian deaths in those cities (the evil effect), not from destruction of any traditionally legitimate military target or even of unavoidable civilian deaths in connection with the destruction of a military target.

Now certainly these criteria for just conduct in war are not capable of precise measurement by the objective observer. Some deal with interior motives; others concern estimates of the probability of various outcomes about which there may always be disagreement. But their purpose is to provide not a basis for judgement on others' acts, but some standards for internalized restraint that might be more effective than the situation arising where wars are judged solely by their causes and ends, when it is so easy to consider oneself wronged.

Using the preceding criteria, let us suppose that nuclear weapons per se are not branded as immoral even though many of their potential uses might be so characterized. The <u>initial</u> use of nuclear weapons seems imprudent to me under most circumstances, given the substantial, though not universal, consensus on the difficulty of maintaining limits in their use once the first have been used. Most analysts of limited tactical or central war now agree that the nuclear vs. non-nuclear threshold is a terribly important and salient one; that given the general absence in the world of either the experience of limited nuclear war or a fully articulated military and political doctrine it is hard to have much confidence in the stability of limits above that threshold.[18] Hence the threshold should be crossed only with great reluctance.

Furthermore, the <u>initial</u> use of nuclear weapons may, by the above criteria, be immoral because the uncertain consequences make calculation of the rule of proportionality into almost sheer guesswork. One possible exception concerns the hypothetical circumstance when a nation's political leaders obtained information, with very high confidence, that their enemy was about to attack them with nuclear weapons. Perhaps then a preemptive strike would be proper if there were no alternative, such as negotiation or delivery of warning. But the conditions for such a strike -- very high confidence in one's intelligence, and the absence of alternatives, are so stringent as to make the scenario border on fantasy. A somewhat more plausible situation applies to the defense of Western Europe. It is possible that European conventional defense is or could become so weak as to offer no reasonable prospect for success against a full Eastern attack, thus posing a very ugly dilemma. But despite the arguments either way, it still is not clear to me that the rich Atlantic nations need allow themselves to become so inferior in conventional forces that they would be faced with such a choice.

The first use of nuclear weapons, when that first use was directed deliberately against civilian targets, would pretty clearly be ruled out by the above criteria. Thus a Soviet conventional attack on Western Europe could not be met directly with an American countercity retaliatory strike. But of course that is not very likely anyway. By a fairly clear extension of the criteria enumerated above, it would also seem that being the first to escalate use of nuclear weapons onto civilian targets would also be ruled out. That is, neither a tactical nuclear attack in Europe nor a counterforce attack on the United States could be met with countercity retaliation. The hardest question is whether nuclear weapons may be used against cities even in retaliation for an attack against cities. The above criteria would seem to rule no on that too, except perhaps in a restrained bargaining fashion. Certainly a merely vengeful retaliation -- to insure that, if one's own society cannot survive their's won't _either_ -- would be indefensible.

Some observers say that the American government need not really _intend_ to carry out such an act of retaliation, that it would be enough to threaten it. Thus a President might say loudly and boldly that he would order retaliation on enemy cities while quietly inside himself rejecting the idea. Bluffs may work in some kinds of interpersonal and international behavior, but in this instance how would a leader convey such a threat, in a credible manner, to the enemy? He would have to repeat it over and over (as both American and Russian leaders have done), and his subordinates would have to be kept completely in the dark as to his real intentions. He could never confide to anyone his unwillingness to carry out the threat because of the extreme importance of keeping his hesitations secret from the enemy. He would have to make the threat a firm national declaratory policy, and so would be responsible for the political climate that would influence his successor, if not himself, to mean the threat as well as to state it. Furthermore, he would have to take all the preliminary steps for carrying out the threat just as though he intended to do it. It would be extremely difficult to keep full control over the decision so that should the country ever actually be attacked, no subordinate would give the firing signal in the protracted absence of the expected word from his commander-in-chief. Thus this particular kind of bluff just does not seem available to the head of a government.

The moral evaluation of strategic postures poses innumerable difficult and divisive issues. I would not want an American President to impose his own moral values (even if they were also mine) on a populace that did not share them. While there are serious difficulties with the concept, the notion that in some sense a democratic leader must remain an "agent" of his people seems persuasive.

In any case, the proposal for a countercombatant strategy need not rest solely on an argument that it is immoral deliberately to kill civilians. One might simply agree that there is something brutal about the readiness with which most of us have accepted a countercity retaliatory posture up to now. It would be still more brutalizing to fight, and even to win, such a war. We cannot completely forget to ask what kind of people we would be in the end. Or at bottom, utilization in war of such a rule of restraint as I have suggested should make it likely that antagonisms would escalate more slowly, and hence less readily destroy a climate conducive to making peace. If accepted also by the Russians, and ultimately by the Chinese, it would offer the prospect, quite aside from moral considerations, of saving a great many civilian lives on the American side.

A COUNTERCOMBATANT STRATEGY AND ARMS CONTROL

In conclusion, consider briefly some implications of an American countercombatant strategy for weapons procurement, arms control, and international politics generally.

1. Some readers may believe that in the present -- and foreseeable -- balance of terror the probability of central nuclear war is really almost zero. They may thus oppose any suggestion that modes of controlling nuclear war be developed, for fear that leaders would then become readier to resort to it. But I fear that the chances of nuclear war over the next decades, whether from inadvertence, crisis desperation, third-party catalyzation, technological breakthrough for one side, in fact are not trivial whatever we do. Nor do I think that implementing my strategy would make nuclear war attractive. If it did lead to any small increase in the chances of war at all, that liability would be more than offset by markedly reducing the chances that any nuclear exchange would lead to the death of hundreds of millions of civilians.

2. The strategy would not in itself exacerbate American relations with our European allies. Indeed, insofar as it would

eliminate Russian capabilities to expand in Europe in any postwar world, our allies might well welcome the change.

3. It is fully compatible with improved American relations with China or Eastern Europe. For obvious reasons the Eastern European states might welcome it even more than would those of Western Europe. Not only is it compatible with better Sino-American relations, an American distinction between China and Russia as military enemies is critical to it. There could be no automatic American strike against China in the event of a Soviet attack on us; a militarily powerful China must be left unharmed to constitute the threat it can suggest most plausibly -- against the state with which it shares thousands of miles of border.

4. It is not incompatible with further efforts to develop invulnerable deterrents on both sides. While there could be some advantages to having a real counterforce capability in the traditional sense, such advantages would surely be cancelled out by the instabilities that would be threatened, and any way strategic retaliatory vehicles are not necessarily the primary targets under a countercombatant strategy. Insofar as a countercombatant strategy might be perceived as additionally destabilizing in a world of vulnerable deterrents, the development of secure retaliatory forces, for example submarine-launched missiles, is to be welcomed. Probably the absence of such

capabilities, on the Russian side, contributed seriously to the non-acceptance of McNamara's Ann Arbor version of "no-cities."

5. On the above reasoning, it is basically not incompatible with ABM systems designed to protect missile-launching sites. It is, however, incompatible with good area-defense ABM systems for protecting vast territories. Attempted Soviet acquisition of such a broad-capability ABM would require strenuous American counter-efforts, or a negotiated and verifiable agreement to refrain.

6. Similarly, the strategy is not incompatible with civil defense systems designed to reduce civilian deaths from heat, blast, or fallout, since civilian deaths are assumed to be essentially superfluous to Russian leaders' calculations of first-strike attractiveness so long as destruction of their military capability is assured. But any widespread Russian effort to put military bases (other than missile sites) or armament plants in secure underground locations would be entirely incompatible. Arms control agreements designed to prohibit such steps might well become of high priority.

7. It is not incompatible with the further development of small, clean, high-accuracy weapons which would cause less collateral damage. It is important that it not require such new weapons as to bring a new round of arms acquisition or pressure

on the Test Ban.

8. For implementing a countercombatant strategy, MIRV is neither a particular threat nor a "no-MIRV" agreement particularly desirable. Insofar as it was necessary for penetrating an ABM system MIRV could be desirable, though an inaccurate MIRV system, capable of penetrating but not discriminating in its targets, would be highly undesirable.

9. A countercombatant strategy would not particularly benefit from a Soviet-American freeze on the number of nuclear delivery vehicles acquired by each side, nor from a negotiated reduction in the number of vehicles. While such action might be beneficial on other grounds, it would be undesirable if it diminished American or Soviet confidence in the ability of the United States very greatly to reduce Soviet war-fighting ability.

10. Because the strategy contemplates striking relatively many targets, it might require the United States to have more missiles capable of surviving a Soviet first strike than does the present assured destruction strategy. This could raise problems of risk, expense, and arms race phenomena. The gravity of this reservation depends partly on technical computations of requirements, and partly on decisions about just how much countercombatant destruction would be required for a satisfactory deterrent. No answer is possible here, but the

mix of weapons and countermeasures is important. For example, with MIRV but not area ABMs, the existing number of missiles could hit a great many targets. MIRVs with small warheads might indeed be much more effective in this role than if used solely against hardened missile sites. Moreover, even if a countercombatant strategy should not be practicable now, it may become so in the future, especially if we want it. Later technology, if we try to develop in that direction, might make feasible certain kinds of offense and defense that would favor the strategy.

11. A countercombatant strategy does not intrinsically require Soviet reciprocation, though there would be major advantages, perhaps to both sides, if the Russians should indicate both receipt of our message of intent and a willingness to do likewise.

12. The strategy is fully compatible with a declaratory American policy of no first use of nuclear weapons.

13. It assumes some continued restraint in general American behavior in international politics, avoiding situations where a Soviet failure to initiate nuclear war in response to American acts would result in a really major shift in the status quo against the U.S.S.R. or its rulers. In this context, of the phrase "peaceful coexistence" the second half needs to be emphasized as much as the first.

NOTES

[1] Statement of Secretary of Defense Robert S. McNamara Before the Senate Armed Services Committee on the Fiscal Years 1969-73 Defense Program and 1969 Defense Budget (Washington: U.S. Government Printing Office, 1968), p. 73.

[2] When I first drafted this paper I thought its proposal was quite new, though not without partial precedents. Since then, however, I discovered an important paper by Arthur Lee Burns that, though differing in a number of major respects, is strikingly similar in others and in basic outlook. See his "Ethics and Deterrence: A Nuclear Balance Without Hostage Cities," Adelphi Papers No. 69 (London: Institute for Strategic Studies, July 1970).

[3] Though see the Fourth Hague Convention of 1907 which, although still concerned essentially with land warfare, clearly forbade the bombardment of "undefended places."

[4] George Quester, Deterrence Before Hiroshima (New York: Wiley, 1966).

[5] Quoted in Robert E. Osgood and Robert W. Tucker, Force, Order, and Justice (Baltimore: Johns Hopkins Press, 1967), p. 217.

[6] Speech delivered at the commencement exercises of the University of Michigan at Ann Arbor, June 16, 1962.

[7] Thomas C. Schelling, Arms and Influence (New Haven: Yale University Press, 1967), pp. 193, passim.

[8] Another reason--and one that would have to be taken into account in any effort to implement the strategy offered in this paper--is simple organizational inertia and bureaucratic routine. In the Cuban missile crisis SAC bombers were dispersed to 40 civilian airports near cities around the United States because that is what the standard operating procedures called for. Thus, without the intention of Kennedy or McNamara, "the 'no cities' doctrine was laid to rest" for the duration of the crisis, when it was needed most. See Graham T. Allison, Essence of Decision (Boston: Little Brown, 1971), p. 139.

[9] See McNamara's 1968 Statement, op. cit., p. 50. Earlier statements had set the figures somewhat higher.

[10] On Thermonuclear War (Princeton: Princeton University Press, 1960, pp. 142-43.

[11] It appears that the term "countervalue" was originally devised to cover whatever the enemy was assumed to value, and hence could have been applied to the targets I suggest here.

Common usage now, however, is to use it synonymously with counterpopulation or countercity.

[12] A target hardened to withstand 60 psi in overpressure will be destroyed by a 20 kiloton bomb striking within a quarter of a mile of it; that same bomb can destroy frame houses up to about 1 3/4 miles away, but not much further. United States Atomic Energy Commission, The Effects of Nuclear Weapons (Washington: U.S. Government Printing Office, 1962).

[13] While critics on the left have frequently attacked the doctrine of assured destruction as calling for damage far in excess of deterrence requirements, several right-of-center critics have denounced its usual interpretation as insufficient for deterrence. They emphasize the Soviet government's presumed evacuation and civil defense capability. Such critics might well find a concentration on military targets as described here to be in fact more plausible as a restraint on Soviet aggressive acts. For doubts about the sufficiency of current targeting see Eugene P. Wigner, "The Myth of 'Assured Destruction,'" Congressional Record, October 13, 1971, p. E10744.

[14] See Johan J. Holst, "Missile Defense, the Soviet Union, and the Arms Race," in Holst and William J. Schneider, Why ABM? (New York: Pergamon, 1969), p. 175.

[15] See Lynn H. Miller, "The Contemporary Significance of the Doctrine of the Just War," World Politics, 16, 2 (January 1964), pp. 254-86.

[16] Robert W. Tucker, The Just War (Baltimore: Johns Hopkins Press, 1960).

[17] Some version of this position is still typical of most "mainstream" thinkers in both Protestant and Catholic churches. See, for example, Joseph McKenna, "Ethics and War: A Catholic Viewpoint," American Political Science Review, 54, 3 (September 1960), pp. 647-58; William V. O'Brien, War and/or Survival (Garden City: Doubleday, 1969); Paul Ramsey, War and the Christian Conscience (Durham, N.C.: Duke University Press, 1961), and The Just War (New York: Scribners, 1968), and Robert Tucker's sections in Osgood and Tucker, op. cit.

[18] The principal recent dissenting voice is probably that of Bernard Brodie, Escalation and the Nuclear Option (Princeton: Princeton University Press, 1966).

I am indebted to a number of people, including Frederick Barghoorn, Betty Hanson, Paul Ramsey, Paul Schratz, Max Singer, Bruce Stinebrickner, Raymond Tanter, H. Bradford Westerfield and Paul Wolfowitz for comments. Of course I am responsible for the results.

Part III

ARMS AND CONFLICT CONTROL

SOME COMPLEXITIES OF ARMS CONTROL PLANNING

K. Wayne Smith and J. N. Merritt

(The following is based on a transcript of remarks by K. Wayne Smith and Jack N. Merritt at the University of Chicago, Center for Policy Study, Arms Control and Foreign Policy Seminar, December 4, 1971).

Over the last few years there has been an increasing demand for a reduction in United States defense spending. Those who want major reductions have argued that there are more urgent and legitimate domestic claims on national resources for housing, welfare, health, education and the environment. Furthermore, Vietnam has compounded the issue bringing forth opposition to the U.S. military establishment.

Those who call for major unilateral reductions in U.S. defense spending often argue that this will set an example which will break the chain of action and reaction between the U.S. and the U.S.S.R.; that U.S. reductions will produce a similar response in the Soviet Union. Others argue that military power is increasingly irrelevant in any case, and that failure of the Soviet Union to follow our lead is of little practical consequence. Finally, some argue that the impetus for the continuing growth of defense spending comes primarily from a military industrial complex which creates its own raison d'etre by developing ever more complex and expensive armaments to respond to only assumed, or, at best, ill-perceived threats.

The basic premises in these arguments cannot be substantiated by either history or experience. There is no evidence that a unilateral reduction in United States military power will call forth a similar reduction in the Soviet military establishment. Indeed, there are serious questions about the degree of real correlation between U.S. force developments and those of the Soviet Union. Their decisions -- like those of the United States -- are dictated at least as much by internal criteria having to do with their own intentions, interests, objectives, and bureaucratic structure as they are by the external threat and force development of others. As for the disutility of military force in the modern world, this issue is far more complex than the simple notion that a nuclear war is all but "unthinkable." Military power has a demonstrable impact on world affairs as recent events strongly attest.

Finally, that a military-industrial complex fuels the "arms race" can be viewed, in large part, as simply a disavowal of the legitimacy of opposing views on the first two issues. Many believe that the electorate of the United States still considers the security of the nation important and that the electorate and their elected representatives support sufficiently large military forces to insure security against all reasonable threats.

It is not our purpose, however, to explore in depth the counter arguments. Rather, it is simply our premise that the differing views on how much to spend on defense and the potentially catastrophic penalties for miscalculation dictate that, for the foreseeable future, significant limitations on arms must come through arms limitation negotiations rather than unilateral actions. If one can accept that stability depends largely on a perception of shared security then classical models of arms negotiations towards equitable arrangements are the only realistic course immediately available for arms control. The design and negotiation of arms control agreements is, however, complicated by a variety of factors including some of the differing perceptions which argue against unilateral arms reductions in the first instance.

This paper discusses some of the major aspects of formal arms control negotiations and outlines some of the difficulties arising from differing but legitimate views.

The arms control record is spotty. Recently some progress has been made in controlling biological and chemical weapons, in restraining the proliferation of nuclear weapons, and in limiting the locations where nuclear weapons might be deployed. But, at best, progress on arms control has been modest and uneven over the years.

The uneven pace gives rise to the central question: Why have some arms control measures been successful and why have others failed? The answer to this question is neither simple nor easily determined since it involves all of the differing perceptions of effects on the political and security interests of the negotiating parties.

THE CONDITIONS FOR NEGOTIATIONS

Arguing that limited agreements are easier to negotiate than comprehensive ones and simple agreements easier than complex ones, while true, doesn't get one very far. For example, general and complete disarmament is a simple concept but so comprehensive in operational terms that it is not even a very useful conceptual goal. Stopping the deployments of ABMs is a much more limited measure but is still far from a simple problem. Even such limited measures present enormous problems not only of definitions but of assessing the strategic implications.

Certainly part of the answer is the fact that negotiations are easier where vital interests are not directly involved. Thus, the banning of weapons of mass destruction from the seabeds was possible since no nation had, or had plans for deploying, strategic weapons on the ocean floor. Similarly, the non-proliferation treaty has attracted many countries but, for a variety of reasons, the vast majority were not likely to produce nuclear weapons in any event. This is not to denigrate the importance of these measures. It simply demonstrates the importance of vital interests and the value of trying to establish controls before the fact.

There is also the matter of who is involved in the negotiations. Negotiations conducted between equals are much more likely to produce agreements than negotiations between unequals, particularly in negotiations regarding strategic weapons. There were no strategic arms negotiations between the United States and the U.S.S.R. in the 1960s because there was no real basis for negotiations. The disparity in the strategic strength of the two countries was too great, too obvious, to permit serious negotiation where such fundamental interests are involved. Strategic Arms Limitation Talks (SALT) were not possible until the U.S.S.R. had the prospect of some sort of parity to carry into the negotiations and the U.S. had determined that parity was acceptable. Similarly, reductions of conventional forces in Europe were inconceivable so long as the U.S.S.R. had a seemingly overwhelming advantage. But, as a clearer perception of the balance in Europe has developed (and, perhaps, as the Soviets have found it increasingly difficult to maintain large forces in Europe while simultaneously putting forces on the Chinese border) conventional force reductions are being viewed with more interest. Again these are some of the broader conditions of arms control but they reflect important, practical problems associated with initiating arms control discussions.

ACCEPTABLE AGREEMENTS

We start with the important premise that while stabilizing arms control agreements are desirable, not all arms control agreements meet this requirement. To the contrary, almost any agreement is potentially destabilizing if it is not constructed with care to forestall both long and short term unilateral advantage. To assess the relative benefits and liabilities is the principal task of analysis. Thus, arms control should be approached with the following basic questions:

(1) What are our goals?

(2) Will an agreement help achieve those goals?

(3) Do we understand the elements of the stability we seek?

What ancillary results can be expected?

Failure to consider proposals in these terms not only runs the risk that the agreement will be destabilizing rather than stabilizing, it also risks promising more than can realistically be delivered with consequent disappointment, distrust and decision born of frustration. Indeed, abortive negotiations are more likely to set the cause of arms control back than to advance it.

In taking a broadly analytical approach to arms control the U.S. has been faced with major differences of view and conflicts in terms of:

--the military balance;

--technology, weapons procurement, and economics;

--policing the agreement;

--negotiations, diplomacy and politics.

Each of these aspects is discussed briefly below, drawing upon our experience with (1) Mutual and Balanced Force Reductions in Europe (MBFR) which is still in the preliminary analysis state; (2) SALT, which is now in the active negotiation stage; and, (3) a comprehensive nuclear test ban, a proposal which has had much consideration and discussion but has seen little progress either in terms of analysis or negotiation.

MBFR, SALT and a comprehensive test ban provide examples

of three distinctly different (although sometimes overlapping) sets of problems confronting arms control negotiations: (1) In MBFR there is the question of whether an agreement will help achieve the goals we seek. What is the existing balance of forces in Europe and how can limitations be designed which are more stabilizing? Are simple reduction formulae adequate or can we perhaps reduce the offense and enhance the defense? (2) In SALT there is the question of ancillary effects on our strategic forces and their capabilities. There are widely diverging views as to what role strategic weapons can play. What are the elements of a strategic balance? Are there credible scenarios involving limited use of strategic nuclear weapons? Is there some level of assured destruction which would deter the Soviets in all instances? (3) In the case of a comprehensive nuclear test ban there is the question of what goals are being sought. Are we seeking arms control for its own sake? Would a test ban force competition into areas with more destabilizing effects on the strategic balance or cause uncertainty about the reliability of one's own deterrent?

THE MILITARY BALANCE

The starting point for any arms control discussion must be an objective appraisal of the existing and subsequent military balance. This appraisal, however, involves differing perceptions not only of existing and future force capabilities but also of the proper missions and expectations for various forces.

Force Reductions in Europe

The debate over relative conventional force capabilities in Europe has gone on for years. On the one hand there is the view that NATO is hopelessly outmanned and that NATO's some 29 divisions available at the start of mobilization (M-day) would be quickly overwhelmed by the more than 80 Warsaw Pact European-oriented divisions. On the other hand, it is argued that there exists a rough balance in the Center Region in terms of real military capabilities. Those holding this view point to such data as contained in the following table which displays the ratio of Pact to NATO strength in selected elements.

Element	Approximate ratio Pact:NATO
Men	1:1
Tanks	2:1
Anti-Tank Weapons	2:3
Aircraft	1:1
Artillery	1:1

And, there is a range of views in between. Such differing perceptions have enormous implications for the structuring of an arms control proposal. If the conventional balance is overwhelmingly in favor of the Warsaw Pact, and if, after 25 years, there has been no redress in the balance by NATO, then there is serious question as to whether there is anything to be lost in a mutual force reduction -- or a unilateral reduction for that matter. In fact, it can be argued that even small monetary savings from reduced NATO forces and the political value of an agreement are worth more than the military loss. In other words, what difference is there between holding the Pact for 5 days rather than 3 days? The end result is the same.

On the other hand, if one views the absence of armed conflict in Europe over the last 25 years as being a function of deterrence -- deterrence based in part, at least, on a rough balance of forces -- then it can be argued that mutual force reductions should be made only if careful study indicates that the resulting balance is not destabilizing. If one accepts this premise, the important questions are: (1) What is the threshold below which we risk having no defense; and (2) What kinds of reductions are truly stabilizing?

If it is agreed that NATO has a credible conventional deterrent then there is some breakpoint, some irreducible minimum below which this would be lost. With the frontage which NATO must defend being fixed, force reductions at some point negate whatever advantages might accrue to a defender without necessarily affecting the advantage of initiative accorded the attacker.

What are the most stabilizing solutions? The idea of force

reductions on both sides may be a somewhat simplistic concept. Many tend to equate "arms reductions" with "arms control" when they should be approaching the issue in a more conceptual way. If increased stability is the goal then the U.S. should consider trying to reduce attack capability while increasing defense capability. The problem is that general purpose forces are not as easily categorized in these terms. A tank is an offensive weapon, but a tank also defends against another tank. Unfortunately there is, at present, only a limited understanding of the role of general purpose forces and the key variables involved in this offensive and defensive effectiveness. (Indeed, SALT with a large body of accepted premises pertaining to strategic forces is relatively simple compared to MBFR with the complexities and uncertainties related to General Purpose forces.) But looking at the problem in this fashion suggests some interesting approaches: Are not static positions, i.e., barriers and fortifications purely defensive, and stabilizing, contributions to military capability? Can we look at the major offensive weapons systems which are most threatening and determine thresholds below which an attacker would not have reasonable confidence of success? Are there ways to mobilize and move forces more rapidly? There is some historical precedent for allowing observers in Eastern Europe. Would not such an agreement contribute to warning and stability, perhaps more than reductions?

As one looks at the problem on a broader scale the possibilities become more obvious but so do some of the limitations. For example: No one really knows what force ratios are needed to attack or what level of confidence the Warsaw Pact might seek; a defensive line across Europe poses serious political problems; and so on. Nevertheless, conceptualizing the problem is more likely to produce hoped for results than simple reduction plans.

Strategic Arms Limits

The question of what we expect of our forces -- what criterion are used for measuring adequacy -- also must be faced. Can we look at strategic forces only in terms of assured destruction calculations? That is, are there enough surviving forces to insure our capability to destroy a significant portion of the population and industry of the U.S.S.R. even after U.S. forces had sustained a major attack. The current conventional wisdom is that as long as we have an assured destruction capability anything else is excess to our needs for deterrence. Some argue, however, that this is overly simplistic and that the strategic force equation is much more complex. Among other things, they contend that: (1) strategic forces must be able to produce a relative advantage in war or the credibility of the deterrence is weakened; (2) any difference between first strike and second strike, however slight, is necessarily destabilizing even if a significant second-strike capability exists; and (3) forces are needed for roles and targets other than cities, e.g., military targets and limited exchanges.

Forces adequate for assured destruction are not necessarily adequate for these other missions or at least have only limited alternative capabilities since they may lack the necessary numbers, accuracy, yield or command and control capability. For example, depending upon the ABM defenses encountered it would require only about 200-300 Minuteman missiles to achieve the generally accepted level of destruction considered adequate for deterrence (i.e., 25%-30% prompt population and 45% industry destruction). This is from a total programmed U.S. strategic missile force as shown below:

Minuteman II	450
Minuteman III	550
Titan	54
Polaris	160
Poseidon	496
TOTAL	1,710

If, however, there is a requirement to attack certain military targets, or to conduct a flexible response to a limited nuclear attack then the nature and numbers of required surviving forces change considerably. And, of course, if one believes that forces must also have a clear capability to produce a relatively favorable outcome at all levels of conflict in the event of strategic nuclear war, then the required forces are many times those needed for simple deterrence.

The key point is that developing arms control agreements depends heavily on the postulated scenarios. These vary widely in intellectual respectability and credibility and each has widely differing implications for arms control.

Not only are there problems in defining missions and capabilities of our own forces, one must also look at those of the Soviets. A major complexity of the SALT negotiations is the differing perceptions of forces between the U.S. and the U.S.S.R.

Soviet and U.S. forces are importantly different:

--Soviet missiles have large payloads and are relatively simple;

--U.S. missiles are smaller and more sophisticated in terms of technology;

--The U.S. maintains a sizeable strategic bomber force;

--The Soviets have few intercontinental bombers;

--A historical component of the U.S. force involves nuclear capable aircraft stationed on the continent of Europe;

--The Soviets have several hundred MR/IRBMs which can

strike Western Europe;

--The Soviets have a deployed ABM system;

--The United States does not yet have an operational ABM.

Throughout the SALT negotiations the United States has had major concerns about the Soviet SS-9 and its potential for destroying Minuteman in a first strike. Moreover, the U.S. has had no on-going deployment of new offensive systems but has relied instead on qualitatively improving existing systems. Consequently, the U.S. has concentrated on constraining the offense. On the other hand, the U.S.S.R. seemingly is filling out its Moscow defense while they see the U.S. deploying an advanced technology Safeguard system. Thus, they have concentrated on constraining the defense.

Moreover, there are other, broader, perceptual differences. In a bilateral negotiation the relative international positions of the two parties cannot be the same. Thus, the Soviets must be far more concerned than the U.S. (at least in the near term) with the growing Chinese nuclear capability.

A Nuclear Test Ban

Nuclear test bans is an area where we particularly need to review our understanding and our purposes in terms of the military implications. The major arguments for a test ban have generally been:

First, that a test ban is needed to prevent extensive radioactive contamination of the environment. Before the cessation of atmospheric testings by the two major powers this was an important argument. However, after eight years of underground testing there is little evidence that such testing has had appreciable environmental side effects. This is not to say that the cumulative effects over time may not become significant but, to date, there seems to be no threat to the environment.

Second, that cessation of nuclear testing would put major constraints on weapons developments and would be, therefore, highly stabilizing. (Interestingly, the argument is also made that weapons development has reached such a sophisticated plateau that further testing is marginally of little value and can, therefore, be foregone.)

And third, that a cessation in testing would bring significant economic gains.

For a number of years attention has focused on the verification aspects of a test ban without carefully considering other questions of military stability. But the key question is whether a ban on testing would in fact be stabilizing even if verifiable. This raises such questions as: (1) Would nuclear weapons development really be constrained or could development continue in the laboratory without testing; (2) What are the possibilities for shifting competition into non-nuclear areas? This could involve esoteric but non-nuclear kill mechanisms or such things as improved guidance for increased capability of missiles; (3) Is the effect of a test ban asymmetrical between countries having first-strike and second-strike policies? Could a test ban erode our capability to sustain a first-strike and still retaliate?

If the arms race is forced into new and different channels then the problems of predictability become greater. Consequently, conservative estimates often force over-reaction and what was to be more stabilizing can become in fact destabilizing. Moreover, as discussed later, forcing competition into more esoteric technological areas tends to increase costs.

TECHNOLOGY, WEAPONS PROCUREMENT AND ECONOMICS
Technology and Weapons Procurement

Technology not only raises the possibility of applying advanced technology to numerically constrained forces and, thereby, appreciably changing the balance, it also affects the likelihood that an arms agreement will accomplish its stated purpose. While similar, these are not necessarily identical concepts.

A major lesson we have learned in preparing for and negotiating in SALT is that technological constraints are difficult to devise, difficult to agree on, difficult to negotiate and nearly impossible to verify.

Quite understandably there is great reluctance on either side to "mortgage the future" by closing off force improvements which might be available through technology. Moreover, the Soviets are concerned that the United States has a sizeable lead in technology and the U.S. is perennially, and justifiably, uncertain about detection of violations because of the closed nature of Soviet society.

Technology also presents special problems because of its inherent unpredictability. Even assuming that agreement could be reached on a measure limiting technology (for example, we can envisage a restriction on testing ABM missiles) such a provision does not close off a number of other developments which might have potential for the same function. There is also concern about the area of technology which is beyond conjecture and control, the area of the unknown where even the most innocent appearing and apparently domestically-oriented development might eventually have major military implications.

Given that technology is difficult to constrain, it follows then that technology offers an avenue for effective circumvention of an arms control agreement. Beyond normal technological progress and force improvement, the arms race could be channeled into a race for qualitative improvements. Indeed, United States strategic forces have remained relatively constant in numbers of missiles for several years. U.S. force changes have been qualitative, for example, multiple independently targetable re-entry vehicles (MIRVs) for Poseidon and Minuteman, and hardening of silos.

What is not understood is whether a "technological race" is stabilizing or destabilizing. There has been in the past a general feeling that qualitative improvements are not destabilizing. However, such developments as high-yield MIRVs with very good accuracy and improvements in anti-submarine warfare capability obviously threaten retaliatory forces and are destabilizing. And, there is always the spectre of a dramatic breakthrough that will upset the balance.

With such worries about the effects of technology, the major conclusion may well be that any agreement particularly sensitive to technological breakthrough should be of limited duration and not extend beyond the "technological horizon." This prevents technology from becoming a barrier to agreements and buys time to continue the search for ways to reduce sensitivity to technological change.

Another of the lessons learned in SALT is that during active negotiations weapons acquisition decisions are artificially accelerated. In developing a strategic arms limitation proposal, decisions have to be made very far in advance as to whether the United States would ever deem it in our interest to deploy certain new systems, for example, mobile ICBMs. This issue has to be resolved on the basis of our ability to verify a limitation and on the assessment of the future role and survivability of the ICBM force and the other elements of the Triad (i.e., land-based missiles, sea-based missiles and bombers). Decisions on other systems (e.g., hard-site defense and ship-based missiles) are also required far in advance of what would normally be the final decision points.

The net effect of such forced decisions is to increase -- rather than decrease -- the number of systems in research and development. Conceived of and easily justified as a major hedge against uncertainties created by possible cheating, abrogation or technological breakthrough, the simultaneous development of a variety of new systems is a likely outcome of any arms control agreement.

Economics

It is important to understand at the outset that economic considerations operate as an important constraint on arms control. Defense spending has a sizeable impact on the economy and redirection of many billions of dollars can have a disrupting effect. While it should be possible to rechannel productive capacity into other areas, this is not easy in the short run without extraordinary measures. As for the expectations of receiving a "fiscal dividend" from arms control, the above discussion of technology should suggest considerable caution.

For example, there is little reason to expect large monetary savings in the short run from a SALT agreement. As we have pointed out, qualitative limitations are difficult to define and negotiate. This means that while numerical constraints are likely to come out of SALT there is still a vast area for competition in terms of greater capability for each allowed system and there will be great pressures to replace aging systems with new ones (e.g., B-52 with the B-1). For example the qualitative improvements being made to, or currently planned for, U.S. strategic forces amount to around $55 billion. And, over time, these costs are likely to increase significantly.

	Total Cost (Billions)
Minuteman III	$ 5.6
Poseidon	7.0
Safeguard	7.0
SRAM	1.0
ULMs	15.0
B-1	12.0
Air Defense	7.0

Moreover, there may be appreciable costs associated with verification of SALT. It is difficult to estimate the size of these incremental costs but increased satellite surveillance alone might be as much as several hundred million dollars per year.

POLICING ARMS CONTROL

The question of verification or policing an arms control agreement is fundamental and is an equally difficult and contentious area. The problem in verification is not only to "know" if the other side violates the agreement, but to have unambiguous "proof" in order to justify some response to a violation for a public audience, either domestic or international.

Depending on judgments as to how much of a risk can be accepted, verification requirements can virtually eliminate various types of arms control. Certainly, many technological limitations could be devised which would alleviate the problems discussed above, but there is no foolproof way to assure that the other side is adhering to the agreement. Whether or not this is the key issue, the inability to verify a comprehensive nuclear test ban has effectively prevented any such agreement for years.

In the United States there is extensive information in the public domain which would be useful to someone trying to determine the salient features of our forces. Moreover, our future plans and therefore any efforts to circumvent an agreement would be public knowledge and such circumvention could not expect to receive the Congressional approval needed. In the U.S.S.R., however, there is almost no substantive military information in the public domain. Moreover, a decision to circumvent an agreement and/or prepare for a major breakthrough with subsequent abrogation could be made without public knowledge.

If one seeks a high assurance of safety, this asymmetry results in limits on the sorts of agreements which might be devised and in provisions which are asymmetrically intrusive or constraining and tend to be more objectionable to one side than the other, thereby further complicating negotiations. For example, the United States has long considered on-site inspection to be an important method of verification while the U.S.S.R. has long objected. Obviously, since there is such a wealth of information on U.S. forces in the public domain the marginal cost to the U.S. (or advantage to the U.S.S.R.) of on-site inspection in the U.S. is very small. Conversely, the Soviet leaders would see themselves as giving much more at the margin by allowing on-site inspection in the U.S.S.R.

At least one conceptual advance from the SALT preparations has been to develop "collateral" constraints to assist in verification. Where the basic limitation (e.g., numerical limits on ICBMs) is subject to verification uncertainties with respect to ambiguous acts (e.g., major changes in medium range missile silos), then the ambiguous act itself might be prohibited by the agreement. While making the agreement more constraining, this approach also allows verification to be less intrusive.

NEGOTIATIONS, DIPLOMACY, AND POLITICS

Negotiations

It has been said that the SALT negotiations would have already been brought to a successful close had the U.S. been negotiating with the British rather than the Soviets. The inference here is that there exists between the U.S. and the U.K. that quantum of understanding, of essential trust and shared interpretation of the world to make the negotiating process much simpler.

As discussed earlier, national perceptions have an important impact on arms control and perceptions of the elements of the strategic balance are particularly important. An example of conflicting perceptions is the differing views of ABMs which prevailed several years ago. In the United States the relation of strategic offensive forces to strategic defensive forces was well understood but, in Soviet strategic thinking, ABMs were (or appeared to be) considered as defensive only and not directly related to the offense. Over the past several years Russian strategic thinking has become more sophisticated and there is a clearer understanding of the first-strike implications of a population defense. Certainly, SALT would have had only limited possibilities without a roughly equal degree of sophistication in strategic thinking on both sides.

But the environment for successful arms control goes beyond equality of strategic sophistication and rough equality of forces. It also requires an atmosphere of favorable relations between negotiating parties. Of course, to some degree, the climate is a matter of political decision as to whether to respond to this or that situation or whether to allow certain events to go unchallenged. By and large the SALT negotiations have been kept separated from tensions which have developed between the U.S. and U.S.S.R. from time to time over the past two years. But, it is much easier to keep talks going in an adverse atmosphere than it is to start them. In 1968 the U.S. and U.S.S.R. were to have begun strategic arms limitation discussions. On the eve of those discussions the Czechoslovakian invasion took place setting back negotiations more than a year. And, in 1961 French nuclear testing was alleged by the U.S.S.R. as the principal reason for the cessation of the testing moratorium.

Diplomacy

It is equally important to recognize that bilateral negotiations are not conducted in a vacuum. The relationship between the negotiating parties and their allies and adversaries is critically involved. When bilateral negotiations begin there is an implicit agreement to allow the rest of the world a relative advantage. In SALT, if the U.S. and U.S.S.R. agree to stop deploying strategic arms, the Chinese are given the opportunity to improve their relative position. If ABMs are limited or banned, France and the U.K. avoid possible degradation of their nuclear strike force. This is a major political decision and a price both sides have implicitly agreed to pay.

Another major relationship has to do with the perceived diplomatic balance as opposed to the carefully drawn military balance. While it is capability not numbers that is important, large numerical differences can and do have important differences with respect to diplomatic leverage. Where allies depend upon our nuclear assurance, large variations in numbers of weapons serve to create uneasiness. This diplomatic aspect of the strategic military balance is too frequently overlooked.

Politics

Domestic politics and institutional pressures on both sides also serve to complicate the development and negotiation of an arms control agreement.

In the United States, the Arms Control and Disarmament

Agency is charged with furthering the cause of arms control. On the other hand, arms control is traditionally met with suspicion by military and civilians alike in the Department of Defense. Moreover, within the military departments one finds, not surprisingly, differing views. It is no surprise to discover that the Air Force is more willing to impose constraints on submarines than the Navy; the Navy returns the compliment with respect to bombers. And, no one in the military favors significant limits on modernization.

The Soviets have similar problems. The sole mission of the Strategic Rocket Force appears to be to control land-based ICBMs. A freeze on both numbers and modernization would leave the SRF presiding over a graveyard and it is certain there is great internal pressure to prevent this from happening. The more comprehensive agreements, however, are not easily designed and negotiated.

CONCLUSION

The foregoing suggests some of the enormous complexities and pressures the President and his advisors face in any attempt to pursue an arms control agreement. Arms control, rather than being confined to a simple negotiation, is a piece of a process involving force planning and deployment, the totality of international relations, internal economic and social pressures, and institutional bargaining. That arms control, therefore, proceeds at an uneven pace is not surprising. The case of the limited test ban treaty is a perfect example.

In March 1958 the Soviet Union ordered a suspension of nuclear tests, reserving its freedom to resume testing if other states continued. The United States was then in the middle of a test series and did not respond. In August, however, the U.S. proposed negotiations for a treaty and offered to suspend testing for one year if the United Kingdom and the U.S.S.R. did likewise. The Soviet Union did not reply, but just after the U.S. test was completed on October 31, the Soviet Union tested on November 1 and 3. The U.S. then announced that it was relieved of any obligation not to test, but that it would continue the suspension for the time being and hoped the Soviet Union would again suspend testing. There then followed a period of nearly three years in which the U.S. did not test and did not detect Soviet testing. The Soviets, during this period, announced that they would not be the first to resume testing. They resumed in August 1961, before the U.S. but after a number of French tests. In 1963, after further testing by both sides, the U.S. announced a moratorium on atmospheric tests in order to help achieve a treaty; this continued for several months until it was replaced by the Test Ban

Treaty. Underlying this uneven pace were arguments over the security implications, worries about verification, and international and domestic pressures.

However, we can and have learned from our experience over the years. In SALT, particularly, we have found that extensive analysis prior to the negotiations can help prevent the negotiations from leading to a deadlock because both sides have not adopted the same stance. With all of the complexities we have been addressing it is highly unlikely that two sides would have the nearly identical views at the outset necessary to permit negotiations to proceed unhampered. In the early phases even the simplest issues are nearly insurmountable. For example, there are widely differing answers to such questions as "What is a tank?" and, "How can you tell an air defense missile from an ABM?" Moreover, assuming competency on both sides, it is highly unlikely that either side can emerge from the negotiations with an obvious military unilateral advantage. Yet both can be expected to try, at least at the outset.

The exhaustive internal analysis of strategic systems and important issues prior to SALT gave the U.S. the flexibility to rearrange the "building blocks" of an agreement (that is, the various combinations of systems and limitations) in order to reduce the likelihood of early stalemates. That we still have many questions concerning the proper conceptual approach to MBFR reflects the complexity of the subject. SALT has dozens of "building blocks," MBFR has hundreds.

Yet, with all of the analysis, our experience is that the most that can reasonably be expected is to "make haste slowly." Arms control is fundamentally a political matter. Opposing perceptions, vital interests, institutional biases and political pressures create formidable hurdles. It requires time and effort to reconcile opposing internal views and for both parties to make the necessary political decisions to turn a proposal into an equitable and durable agreement. It also requires the will.

Part IV

DECISION-MAKING AND POLITICAL LEADERSHIP

THE PRESIDENT, THE CONGRESS AND
ARMS CONTROL

Adam Yarmolinsky

The story, perhaps apocryphal, has been told before of Secretary McNamara emerging from a long debate with the Joint Chiefs early in 1961, during his first few months in office, to report to President Kennedy, with some satisfaction, that he has persuaded the Chiefs to accept a force level of 1000 rather than 1200 Minuteman missiles. The President scarcely paused before inquiring, "Why not 500?" But the real point of the story is that the number of missiles did not go below 1000. It may be that 1000 was the ideal figure. But the President did not have the staff to pursue the issue in depth, through the convolutions of Pentagon politics as well as of nuclear strategy, while the Secretary of Defense had been positioned into a choice between two numbers, neither of which may have been valid.

It can be argued that adjustments in the force structure are not a matter for amateurs, and that outsiders, even the highest politically responsible officials, should either leave them to the generals, or learn enough to argue with the experts on an equal footing. But in the nature of the situation, even the most indefatigable outsiders cannot be expected to achieve an equal footing with the experts within the military establishment. They cannot build and maintain the kinds of staffs and data banks that would be needed. At the same time, the outsiders inside the Establishment, the Secretary of Defense and his staff, tend to lose perspective because of their close and continuous involvement with military problems.

If the President (or the Congress) finds it difficult to address a specific issue affecting the size and composition of the nuclear forces, how much more difficult it is for them to address questions affecting the size and composition of the non-nuclear forces. The relationship between nuclear forces and foreign policy is a good deal less complex than the relationship between non-nuclear forces and foreign policy. The nuclear threshold is so high that nuclear forces are relevant to a very small range of foreign policy considerations, while the possible uses of non-nuclear forces, whether as deterrents or in actual combat situations, are enormous.

The principal determinant of the size and shape of the United States military budget and force structure ought to be the military requirements, nuclear and non-nuclear, of United States foreign policy. To translate foreign policy goals and objectives into military hardware and manpower has been the presumed, and sometimes even the stated aim of Secretaries of Defense since the office was created almost a quarter of a century ago. Some have succeeded better than others. Success has tended to favor those who presided over an expanding military establishment, and to elude those who presided over a contracting one. This is hardly surprising, since the Secretary of Defense, in order to maintain his own position, must be responsive, in some degree, to the needs of the establishment itself. An organization as enormous, as powerful, and as extensive as the United States military establishment has a considerable dynamic, and contraction is not its natural mode.

The President and the Congress, the two forces within the American system of government to which the military is directly answerable, stand at a greater distance from the military establishment than the Secretary of Defense and his associates, who make up the politically responsible civilian leadership of the establishment. If the military budget and force structure are to be trimmed down, to include only what is needed to carry out United States foreign policy, the President and the Congress may have to do most of the trimming. But the logical connection between a pattern of treaty obligations and an inventory of tactical nuclear weapons (or fighter aircraft, or carrier task force auxiliary vessels) is seldom clear and never simple. The judgements required in shaping and reshaping a military budget and force structure are extraordinarily technical and complex. The avenues of intervention that are or may be available to the President and the Congress are, therefore, critically relevant to their ability to achieve a closer correspondence between foreign policy and military policy, particularly in the descending phase of the cyclical ebb and flow of military spending. Indeed if we could imagine, for the purposes of this paper, that foreign policy is held constant, we might be able to examine the consequences for arms control policies of alternative mechanisms of Presidential and Congressional participation in the arms control process.

At the outset, it must be pointed that the two variables, foreign policy, and arms control policy, are not always clearly distinguishable -- as in other cases involving means and ends. Foreign policy involves the selection of national goals or objectives in the world, some of which the nation may seek to achieve (or defend) in whole or in part, by military means. Foreign policy also involves the decision when and how to invoke military means in specific crisis, or near-crisis situations. Arms control policy involves the choice of military means to be available as instruments of foreign policy in the event that they are thought to be needed. Arms control policy also involves the advance determination as to which military means will be employed.

This kind of contingency planning, if it is rigidly structure, can affect foreign policy in major and disastrous ways, as Barbara Tuchman has pointed out in "The Guns of August." But even the most flexible contingency planning limits the choice available to foreign policy makers. European Starfighter pilots whose planes have been configured for nuclear missions will inevitably not be as well prepared for the alternative contingency of non-nuclear missions. And the decision on whether or not to begin the development of a particular weapons system can, if it becomes known, affect the international climate in ways that may require changes in the foreign policy of the nation taking the arms control decision.

Foreign policy and arms control policy interact because both policies, necessarily and increasingly, involve choices among scarce resources. The limitations on the effectiveness of any nation's policies, outside the geographic area in which it exercises national sovereignty, are increasingly apparent. And the limitations are growing on what weapons a nation can afford to buy, because of their increasing cost and complexity, and what weapons it can afford to use, because of their increasing potential for destruction.

Even the most powerful nation in the world cannot choose foreign policy goals without regard to the limitations on military policy -- as the United States has learned to its sorrow in Vietnam -- and by the same token it cannot consider alternative military means to achieve foreign policy goals without examining the limitations on future foreign policy choices that may result from employing one or another of those means.

Thus the notion of holding foreign policy constant while examining alternative ways for the President and the Congress to affect arms control policies is patently artificial. It becomes somewhat less artificial if we assume, as seems reasonable, that United States foreign policy goals in the seventies have become significantly less ambitious than they were in the fifties and sixties, and that the problem for the President and the Congress is to select military policies that are consistent with these more limited goals, without destroying the ability of the United States to muster the military means to achieve them. It is within this narrowing range of foreign policy choices that military policies must be selected.

Thus far we have been using "military policy" and "arms

control policy" as interchangeable terms. In the broadest sense they are always interchangeable, since the job of the military policy maker is to select from the infinite range of force structures and weapons systems, those that are most appropriate to his goals. He cannot have them all, under any circumstances, and he must control the natural tendency of the military establishment towards automatic and general proliferation.

But where foreign policy goals are being limited, arms control in a more rigorous sense is required if the military servant is not to get out of step with his foreign-policy-making master. And arms control in this sense is important not only to avoid waste of resources, or the provision of means for undesirable military adventures, but also to signal to the national security bureaucracy that their masters really mean what they say about changing foreign policy directions.[1]

For our purposes, it is possible to distinguish two kinds of arms control, substantive and procedural. Substantive arms control is concerned with the size and shape of the force structure, and procedural arms control is concerned with its deployment and rules of engagement. Of course what you buy depends on how you plan to use it, and vice versa. But substantive arms control is more responsive to budgetary constraints, and, therefore more easily achievable.

The central dilemma of Presidential or Congressional arms controls is a budgetary one: Should authority outside the military establishment fix an arbitrary budget ceiling, or should it examine every element of the proposed budget, and make case-by-case decisions without regard to arbitrary or predetermined budget ceilings. To concentrate on the overall spending total is to surrender control over the composition of that total to the military itself. But to focus on the details is to risk losing control over the total, which can grow and grow as new or improved weapons systems are shown to be essential for national security.

The budgetary philosophy of the Kennedy Administration, that the United States could afford all the military spending it needed,[2] did not inevitably result in increased military budgets, but it would have been a good deal more difficult to apply in a period of contracting expenditures. Even a firm opinion that military spending has grown disproportionately to its foreign policy objectives tends to dissipate itself in discussions of the relative merits of two kinds of new main battle tank, or the loiter time requirement for a new fighter-bomber.

On the basis of a limited number of political and economic assumptions about acceptable and unacceptable damage on various sides (Russian, Chinese, U.S.) it is possible to discuss as an issue of military policy how much is enough in nuclear weaponry. The political and economic assumptions on which a non-nuclear force structure can be based are so complex that it is impossible to set them out in a way that will permit rational discussion of the size and shape of the non-nuclear forces, as a question of military policy. In retrospect, the simple two-and-a-half war formula seems as unsatisfactory as its successor, the one-and-a-half war formula. It is just not credible that the United States can maintain a military establishment capable, without major expansion, of fighting one-and-a-half wars, much less two-and-a-half. Yet almost any size of non-nuclear establishment would be big enough to get us so deeply into a conflict situation so that we might have trouble backing off, if we were not immediately successful. Unless one assumes that U.S. non-nuclear forces will never be called on again except to defend our physical frontiers, then the one thing we can be sure of is that we cannot now describe with any degree of accuracy the kind of situation in which they will be used -- even as one of a large number of contingencies.

Given the impossibility of developing an overall foreign policy rationale that will provide specific guidance for non-nuclear military policy, one is thrust back on Lindblom's partial incrementalism[3] as a method of analysis for outside intervention by the President and Congress. As to any proposal from the military for a change in the force structure, or the general budget level, the issue becomes, not what should be the ideal force structure of budget level, but how much should it go up or down as a result of decisions that can be taken in the next budget year.

The size of the United States forces in Europe is perhaps the most dramatic and certainly the most important case in point, because the size of the United States NATO contingent more than anything else determines the overall size of United States non-nuclear forces. The issue here is essentially a political one, going to the confidence of the NATO allies in the United States' commitment to Europe and also to balancing United States forces within NATO against the German contribution, so as not to leave West Germany as the single most powerful military force within the Alliance. And the issue is not the absolute size of the American contingent. If it were significantly smaller than it is now, its political impact would not be significantly reduced. What is important is the fact of change and its incremental influence. Any examination of a proposed new United States force level in Europe must focus on the increment, not on the absolute.

If the President and the Congress should concentrate on incremental issues in non-nuclear arms control, because they cannot hope to develop an adequate foreign policy rationale that will translate into an overall non-nuclear military policy, there is also some justification for an incremental approach to the control of nuclear weapons. The balance of nuclear terror is now reasonably stable. The danger is in continuing rounds of escalation, because each nuclear power is pursuing a kind of "worst case" approach to the analysis of the other powers' capabilities. There is also the danger of a stalemate resulting from unwillingness to consider seriously mutual U.S. - U.S.S.R. reductions. Because research and development is so slow a process in nuclear weaponry, there is time enough to negotiate on modernization of existing weaponry without escalation in overall destructive power. Existing systems can be left in place until they become obsolete or dangerously vulnerable. An incremental approach, therefore, can reconcile the pressures of new technology with the needs of arms control; it can keep military expansion in check while avoiding major confrontations between the military establishment and outside political authorities.

Effective participation by President and Congress in incremental decision-making about arms control has several prerequisites: timeliness, discipline, objectivity, and information.

The choice among increments puts a great deal of weight on

how the options are selected. If the decision maker feels free to scrap the entire existing system and start over, it doesn't matter so much at what point he first enters the decision-making process. But if he can only move it a bit in one direction or another, it matters very much what decisions are taken by default, and what implicit major premises are established before he comes along.

The massiveness and complexity of the military budget and force structure make early outside involvement in proposed changes even more difficult. Significant force structure changes inevitably have budgetary consequences, and tend to occur as a function of the annual budgetary cycle. That cycle in turn involves so many levels of review within the Pentagon that by the time the proposed budget reaches the President and his staff, and the Congress and theirs, its form is largely fixed, and many of the original choices have been obscured -- or compromised.

In the Kennedy Administration, issues were focussed by two instruments, both devised by Secretary McNamara, the Draft Presidential Memorandum, and the Secretary's Military Posture Statement. The draft Presidential Memoranda, each covering one of the six or eight major program packages, laid out rationales, costs and options for the President. The Memoranda always left the Defense Department for the White House in draft form, so that no final documents would be prepared until the Presidential input had been incorporated. Similarly, the Posture Statement attempted to put before the Congress the major issues underlying the budget. Both documents, however, depend on the will and ability of the Secretary of Defense to use them for genuine analytical purposes, rather than as political broadsides. And in the nature of their timing in the decisional process, they tend to come too late.

Efforts by the Nixon Administration, through the Assistant for National Security Affairs, to initiate formal White House and interagency reviews much earlier in the process should go a long way towards permitting more timely Presidential intervention. But timely intervention require more than formal relationships. In order to catch the critical moment for incremental decision-making, a President must be able to take advantage of informal staff-level communications between the White House and the Pentagon, including the exchange of internal staff papers, so as to maintain a continuous awareness of the climate of thinking on both sides of the Potomac.

This sort of informal communication is in some ways easier and in other ways more difficult to achieve between the Pentagon and the Congress. Senior military officers have traditionally been in close and continuing communication with senior members of the Armed Services Committees and committee staff. But these exchanges of views and information have traditionally been focussed on the common interests of the two interest groups in expanding the military establishment, without particular regard to foreign policy objectives, but only to the dynamic imperatives of bureaucracy -- of which more below.

Senators, congressmen and committee staff members interested in making the military a better servant of foreign policy have particular difficulty establishing contact with their opposite numbers, even in the Office of the Secretary of Defense. The periodic breakfast sessions in the Pentagon Blue Room during the Kennedy Administration given by OSD staff members for the young Turks on the House Armed Services Committee had very limited attendance. Ancient jealousies, as well as constitutional doctrine, make it very difficult for members of the executive and legislative branches to meet informally, except on the grounds of short range self-interest, which seldom extends to cover arms control objectives. In fact it may not be possible to achieve more timely involvement of the Congress in military policy making, pointed towards arms control objectives, until that involvement can become both more disciplined and more objective.

In fact the legislative process, as it affects military budget making, is a good deal more disciplined than it is generally given credit for. Compared, for example, to the Rivers and Harbors Bills, the Defense authorization and appropriations are handled as models of decorum. The Rivers and Harbors Bill is annually pushed and pummelled through the Congress until it emerges in a shape scarcely recognizeable as what it was when first introduced by the Administration. Defense legislation, on the other hand, generally emerged virtually unscathed, with few, if any, deletions, to be sure, but with few, if any, additions as well. An eager defense contractor, pressing for the adoption of a new weapons system, can almost never persuade the Congress (much less the White House) to put an add-on in the budget, no matter how many powerful allies he can muster, if he has failed earlier to persuade the authorities within the Pentagon itself.

But on too many occasions the administration has not been able to line up its own troops, or the leadership on Capitol Hill (when it represented the opposite party to the administration) has not been able to organize the opposition. In the Eisenhower Administration, Secretary McElroy had to ask a congressional committee to "hold his feet to the fire" in order to make necessary budget cuts. In the Kennedy and Johnson years, eminently sound proposals for changes in the force structure died a-borning because, in Pentagon parlance, "the building wouldn't stand for them." And during the Nixon Presidency, no efforts to resist administration proposals for new weaponry, however controversial, have captured the effective support of the Democratic leadership.

Undoubtedly one factor that undermines discipline within the executive branch and in the Congress, when their leadership attempts to resist expansionist pressures from the Pentagon, is the provision of the National Defense Act of 1947, as amended, that authorizes the members of the Joint Chiefs of Staff to express their views on national security matters directly to the President and the Congress. This provision for what has been described as statutory insubordination, however, only makes explicit a process that is inevitable in any bureaucratic structure. If the principal permanent officials in the structure disagree with a proposed course of action, they will make their views known, no matter what vows of silence are imposed on them. In fact the built-in differences among the military services and their official spokesmen create some of the best opportunities for movement within the bureaucracy.

But a President whose administration is not responsive to his views cannot hope to shape any bureaucracy, particularly the largest and most powerful one in government, to conform to those views. And while the United States Congress is at a far remove from the standards of party discipline in, say, the British Parliament, a Congressional leadership that cannot command a working majority on defense policy issues can only ratify the proposals of the Executive Branch.

It is particularly difficult for the Congress to achieve either discipline or objectivity on military policy issues because both are so clouded by self-interest. So long as the military establishment continues to be the major purchaser of goods and services within the federal government, individual congressmen and senators will have a powerful interest in maintaining or expanding that purchasing power within their own territory. It is unlikely that this purchasing power will be so far diminished, even if it were to be cut in half, that it will not be the largest piece of federal spending. Perhaps a better route to greater objectivity would be to reduce the dependency of individual firms and of geographic areas on military spending. Defense contractors can be required to diversify into other activities, perhaps at a prescribed minimum rate of diversification, depending on their size. And particular areas can be helped to diversify their economic base, as they are helped now by the Economic Adjustment Adviser in the Pentagon, when a plant or base closing cuts their employment sharply.

Diversification is not easy for defense industry, since the habits of doing business with the Defense Department are very different from the habits of doing business in the commercial market. But there are other fields in which firms that learned to do defense business can do business with other departments of government that are looking for high-technology, limited-volume products -- in education, for example, or in traffic control, or in refuse disposal. Legislation that compels a degree of diversification, on the one hand, and eases the pain of retooling on the other, can reduce the pressure on congressmen, and on the executive branch, to maintain present levels of defense spending without regard to foreign policy objectives. Given the fact that constituents who have defense business will still not want to give it up, the pressure on the Congress to find funds for domestic needs and problems can, over time, complete the task of restoring its objectivity.

These developments are by no means inevitable. The present size of the military budget, although it has been gradually declining over the last few years, in constant dollars, and even more as a percentage of gross national product, may be so large that it will be stuck at about its present level without regard to foreign policy considerations, unless and until a strong president, by a great act of will and persuasion, can unstick it.

An essential ingredient, in any event, of the effort to unstick it is adequate information -- a commodity extraordinarily difficult to come by, even in the relatively simple area of nuclear weapons policy, and on so elementary a question, for example, as the precise number of missiles in the United States inventory. There are two basic ways in which a bureaucracy exerts its influence on the decision-making process: by controlling the flow of manpower through hiring and promotion systems, and by controlling the flow of information. There are some imaginative ways in which the President or the Congress can affect the officer recruitment and promotion policies of the Armed Forces, without doing serious injury to the already wounded morale and esprit of the officer corps. Most of the useful recent suggestions in this field have come from my colleague Morris Janowitz. But there may be even more to be accomplished by clearing up the flow of information.

Because of the extraordinary complexity of information about military budgets and force structures, the selection of facts is even more highly colored than in most situations by the sources from which the facts are obtained. Whether they go to the military importance of a particular overseas base, the vulnerability of an allegedly obsolete weapons system, or the superior qualities of a proposed new system, the original source of the facts has chosen them for a purpose. But by the time they reach the highest levels within the military establishment and emerge into public view, they are enshrined in authority, and the bias of their original source may have been obscured. A good deal of the value of the systems analysis operations introduced on a large scale into the Secretary of Defense's staff work by Secretary McNamara lay in the questioning of factual as well as conceptual assumptions in the underlying staff work done by the military services. Questioning the assertion that NATO was outgunned in conventional forces by the Soviet bloc because the Soviets had more divisions, the Systems Analysis staff asked, just how does a Soviet division compare to a U.S. division, in manpower and firepower -- and it turned out that the two divisions were quite different in size and strength.

The independent information resources available to the President and the Congress are not inconsiderable, however. They include experts in the private sector, who have acquired their expertise within the military establishment, and who are still capable of shrewd appraisal of official information. There is some admittedly very limited opportunity for quasi-independent fact-finding by the National Security Council Staff, and a little more by Congressional committee staffs.

Too little use is made by President and Congress alike of the staff of the Arms Control and Disarmament Agency within the Department of State. There is serious question whether that agency should ever have been organized as quasi-independent entity within the State Department, thus isolated from the decision-making process within the Defense Department, to which the bulk of its energies are addressed. As a quasi-independent entity attached to the Pentagon, it might not have been allowed to function, but in its present location it is rather like the desert flower, blooming unseen by human eye. Its staff resources might well be utilized by the President and also by the Congress as an independent check on the factual assertions of the military establishment. A separate staff agency within the legislative branch, or an arm of the Legislative Reference Service of the General Accounting Office, would not be likely to attract the quality of staff required to engage in useful dialogue with the Pentagon's own analysists. But the ACDA staff could do so, if they only had equal access -- never mind equal time.

One of the problems of equal time is the heavy flow of information from the Pentagon itself. Without attempting to draw lines between information and propaganda, one can venture the generalization that informational material generated by a government agency (or by any group or organization) is not likely intentionally to show the agency in other than a favorable light. The difficulties arise not because information emanating from the Pentagon is more highly colored than information emanating from, say, the Department of Agriculture, but rather because there is so much of it, the great bulk coming from the three military departments, without effective supervision by the Assistant Secretary for Public Affairs in the Office of the Secretary of Defense. A greater degree of restraint on inevitably self-serving activities, like the travelling seminars of the Industrial College of the Armed Forces, or the Joint Civilian Orientation Conference, would reduce the amount of noise in the environment where the President and Congress are trying to hear what is really going on.

All of these considerations -- timeliness, discipline, objectivity and a more adequate flow of information -- assume a basic motivation on the part of the national political leadership to make the military establishment a better servant of United States foreign policy. Americans tend to be careless of national resources, using and neglecting them with equal abandon. The military establishment is the locus of major resources of manpower, skills, and physical plant and equipment that, left to develop in accordance with their own internal laws of growth, would represent a tremendous waste of human energy. There is no more excuse for leaving the shaping of the military establishment to the military bureaucracy in a period of shrinking budgets and force structures than there was in a period of expanding budgets and force structures. In fact there is less reason, because the choices are tougher. They need be made, in the last analysis, by the political leadership of this country, in the Presidency and in the Congress.

NOTES

[1] This argument is spelled out in Graham Allison, Ernest May and Adam Yarmolinsky, "Limits To Intervention," Foreign Affairs, Vol. 48, No. 2 (January 1970), pp. 245-261.

[2] President Kennedy's Special Message to Congress on the Defense Supplemental Budget (Washington, D.C.: U.S. Government Printing Office, March 28, 1961): "Our arms must be adequate to meet our commitments and ensure our security, without being bound by arbitrary budget ceilings."

[3] D. Braybrooke and C. E. Lindblom, A Strategy of Decision (New York: Free Press, 1963).

THE SECRETARY OF DEFENSE AND THE JOINT CHIEFS OF STAFF: THE BUDGETARY PROCESS

Lawrence J. Korb

One of the main focal points of the pressures of the military-industrial complex is assumed to be the annual defense budget process. For it is here that the overall size and distribution of military funds is decided. Those concerned with the influence of the military-industrial complex should be interested in how these decisions are made.

Preparation of the defense budget within the executive branch of government is the prime responsibility of the Secretary of Defense. But, the enormous size and complexity of the Department of Defense (DOD) makes it mandatory that he delegate much of the actual budget making to high ranking military professionals, that he rely on these men for advice on those decisions that he reserves to himself, and that he bargain for their support on the final budget.

The highest ranking military men in DOD are the Joint Chiefs of Staff (JCS). The National Security Act of 1947 and its subsequent modifications established the Chiefs as the principal military advisers to the Secretary and as the military heads of their respective services.

A survey of the literature on the military-industrial complex reveals that top military leaders and high level civilian officials in DOD are always considered to be part of the complex. Therefore, an analysis of the parts played by the JCS and the Secretary of Defense[1] in the defense budget process would seem to be vital to an understanding of the scope of public policy that the military-industrial complex dominates.

Since the creation of DOD in 1947, ten men have held the post of Secretary of Defense. However, four of these men have served less than 18 months. This is too short a time to have a real impact on the budget process.[2] Therefore, this study will analyze the parts played by the Secretary and the JCS only during the tenure of James Forrestal and Louis Johnson from the Truman Administration; Charles Wilson and Neil McElroy from the Eisenhower era; Robert McNamara, who served both Kennedy and Johnson, and Melvin Laird, the present Secretary. All of these men have served at least 18 months in office, and although their number is comparatively small, i.e., six out of ten, their tenure does account for over 80 per cent of the period since 1947.

THE TRUMAN ADMINISTRATION

James Forrestal worked with five military chiefs on two budgetary evolutions, i.e., a supplement to FY 1949 budget and the FY 1950 budget. In each of these evolutions, Forrestal and the Chiefs played similar roles. The Secretary of Defense received definite ceilings from the Bureau of the Budget. He communicated these amounts to the JCS and asked them to produce a budget within those constraints. The Bureau's ceilings were based on a Truman decree that military activities were to have one-third of the total budget.[3]

Because of their concern with the world situation,[4] the JCS submitted budgets well in excess of these ceilings. Each Chief was willing to cut the total package only if the majority of the cuts came from the other services. Their budget submissions to Forrestal were simply the addition of the three individual service budgets. In FY 1949, the Chiefs asked for a $9 billion supplement[5] and in FY 1950 they requested a total budget of $29.4 billion.[6] The ceilings for these budgets were $3 billion and $15 billion, respectively.

Forrestal shared the Chiefs' concern about the world

situation and sought assistance from two outside sources to bridge the gap, but neither helped. He created a board of non-JCS officers to reduce the budget, but it made only a small reduction. He then asked the National Security Council (NSC) and the Secretary of State to provide planning guidance to the JCS to enable them to adjust the world situation to administration ceilings, but both the Council and Secretary of State Marshall refused to become involved in the budget process.

Forrestal finally resorted to personal diplomacy. He summoned the Chiefs to his office and told them that they could not have satisfactory and usable military power under the budgetary limitations, but that, if they would produce a budget in the vicinity of the President's ceilings, he would make further attempts to get the President to raise the limitations somewhat. For him to present the Chiefs' original estimations to the administration would strain both his and their credit. But, if they came close, he could tell the President they were not taking his decision lightly.[7]

The Chiefs cooperated, but the Administration did not. In FY 1949 the Chiefs and Forrestal agreed on a $3.48 billion supplement and in FY 1950 they settled on $16.9.[8] Forrestal took these figures to the White House and made impassioned pleas to the President, but Truman refused to raise the ceilings.

Moreover, the Bureau of the Budget reduced, and Truman affirmed, further cuts of $.38 billion in FY 1949 and $1.68 billion in FY 1950.[9]

Louis Johnson assumed the helm at the Pentagon while Congress was hearing testimony on the FY 1950 budget, and the JCS, were working on a statement of forces and major national requirements that would provide a foundation for a FY 1951 budget of less than $15 billion. Johnson did not wait long to make his impact felt on either budget.

Within a few weeks of his appointment, he boasted to the Senate Subcommittee on Military Appropriations that he could save a billion dollars in DOD by cutting out waste, duplications, and unnecessary civilian employment. Obviously impressed, the Senate gave him authority to reduce expenditures by $434 million.[10]

Within a month after succeeding Forrestal, i.e., on April 23, 1949, and less than a week after the Navy had completed well publicized keel laying ceremonies, Johnson cancelled construction of the 65,000 ton flush deck super carrier, the United States.[11] This was done despite the fact that the Navy already had about $1 billion invested in it and Congress had appropriated funds for the ship for three consecutive fiscal years.

In June, the Budget Director informed the Secretary t

DOD's one-third share for FY 1951 would be $13.5 billion.[12] Without asking the services for its implications on their programs, Johnson then ordered a reduction in FY 1950 expenditures of approximately $1 billion and directed the JCS to agree on a budget for FY 1951 below $13.5 billion or face the prospect of an across the board reduction by him.

On two occasions during the summer of 1949, President Truman met with Johnson and the Chiefs to discuss the adequacy of the ceiling. On both occasions the Secretary and the Chiefs assured the President that $13.5 billion was more than sufficient.

The service budgets, which the Chiefs submitted to Johnson amounted to $13.31 billion. Johnson cut another $120 million from the budget on his own and sent it to the White House. The Bureau of the Budget reduced this figure by another $1 billion and Johnson accepted the final figure of $12.21 billion with very little argument.

THE EISENHOWER ADMINISTRATION

Two of Eisenhower's appointees as Secretary of Defense, Charles Wilson and Neil McElroy, remained in their positions for at least 18 months. The President's third appointee, Thomas Gates, left office with the Republican Administration in January 1960, after only 13 months in office.

Wilson remained in office longer than any previous Secretary, 56 months, and McElroy lasted 26 months. The General Motors executive was responsible for six budgets and the Proctor and Gamble president produced two. Both of the former corporation executives played similar parts in the budget process.

The production of the defense budget within the executive under Eisenhower usually took an entire calendar year.[13] The process was initiated about 18 months before the fiscal year in which the budget was to be effective and 12 months before the budget was to be submitted to the Congress. For example, work on the FY 1957 budget, which was to be submitted to Congress in January 1956 and which would become effective on July 1, 1956, began in January 1955 and lasted until December of that year.

The process was inaugurated in January when the National Security Council (NSC) produced the Basic National Security Policy (BNSP) document. This document was supposed to be a comprehensive statement of American strategic policy and had as one of its main purposes providing guidance for the JCS in their planning for force and weapon levels.

Although the NSC devoted a great deal of time and energy to the drafting of the BNSP, the document was useless for budgeting purposes. Rather than resolving the sharp differences of opinion over what the strategic policy of the United States should be, it glossed over them to make the document acceptable to all parties. It meant all things to all men and settled nothing. For example, the 1958 edition stated that the United States would depend upon the weapons of mass retaliation, but at the same time maintain flexible forces capable of coping with a lesser situation.[14]

After completion, the BNSP was sent to the JCS to serve as a guide for their Joint Strategic Operations Plan (JSOP). The JSOP is a document of many volumes that prescribes the forces that the JCS believe are required to carry out military strategy and national objectives. Because the BNSP could be interpreted in so many ways, each service Chief stressed that portion of the BNSP that enhanced the primary mission of his service. Consequently, the JSOP was really three separate plans added together and called a joint plan.[15]

The completed JSOP was sent to the services in late June. It was supposed to serve as a guide or framework for the individual military departments in the production of their separate budgets.

While the JCS were completing work on the JSOP, the NSC was deciding upon a ceiling for defense expenditures for the next fiscal year. The general rule laid down by the President was that defense expenditures should not exceed 10 per cent of the Gross National Product. The specific ceiling for a given year was obtained by estimating total income, subtracting the projected expenditures of all other government agencies, and then allocating the remainder to defense provided it did not exceed the 10 per cent figure.[16] The object was a balanced budget in every fiscal year, and not even such crises as Sputnik or Suez could alter that objective.[17] Defense expenditures of the Eisenhower Administration always came out to between 9 and 10 per cent of the Gross National Product. The defense ceiling was transmitted to the service Chiefs by the Chairman in mid-summer.

The individual service budgets were submitted to the JCS in September. The JCS were supposed to review the budgets for conformity with the BNSP and the JSOP. The Secretary expected the Chiefs to produce a total defense budget which conformed as closely as possible to the plans but did not exceed the ceiling.

However, the service budgets exceeded the ceiling by an average of 15 per cent.[18] Not once did these JCS budgets come reasonably close to the ceiling. On three occasions the budget requests exceeded the ceiling by more than 20 per cent. The JCS refused to trim the budgets. Because of the vagueness of the BNSP and the ambivalence of the JSOP, the Chiefs could and did justify every item in the service budgets and refused to make reductions on the grounds of national security. Wilson and McElroy, who were in sympathy with the President's desire for a balanced budget,[19] tried a variety of approaches to induce the JCS to scale down these requests. These approaches ranged from direct commands to subtle hints, but always had the same lack of results.

In Eisenhower's first year in office, 1953, Wilson twice directed the JCS to indicate where reductions could be made in the services' $37.82 billion request to bring it under $35 billion. The JCS refused to follow either directive on the grounds that any reductions would increase the danger to national security.[20]

The following year Wilson directed Chairman Arthur Radford to persuade the Chiefs to make the desired reductions. Radford was unable to get the Chiefs to agree on a lower level of expenditures and was forced to make the recommendations for reductions himself.[21]

In 1958, when he was faced with a JCS approved budget of $44.67 billion and an administration ceiling of $38 billion, Secretary Neil McElroy tried a more subtle approach. He asked the Chiefs how they would divide up $38 billion without in any way implying that this was the amount they approved. The JCS discussed this hypothetical question briefly and then informed McElroy that they unanimously agreed on how to split up $34 billion, but each Chief felt he needed the additional $4 billion for his own service and could not voluntarily give it away to another military department.[22] In essence, they were telling the Secretary that they needed the entire $45 billion.

In 1959, Eisenhower intervened personally by inviting the Chiefs to a stag dinner at the White House. During the dinner, the former General gave the Chiefs a pep talk on the great need for more cooperation on their part with the Secretary in connection with the budget. Only Chairman Twining was receptive. The service Chiefs refused to change their behavior.[23]

So little did the JCS have to do with the final defense budget that on many occasions they never even considered such important questions as the size of the Army, the number of aircraft carriers or the amount of deterrent forces.[24] The job of making such important decisions, i.e., where to make the major reductions was left to OSD and the President. Wilson and McElroy often ordered across the board reductions. DOD's comptroller Wilfred McNeil made most of the detailed decisions.[25]

If the Director of the Bureau of Budget objected to any of OSD decisions, it had to appeal to the President. Such appeals and Eisenhower's own desires led to his making about 15 major budgetary decisions annually.[26]

KENNEDY AND JOHNSON - "THE McNAMARA YEARS"

Before assuming the Presidency, John Kennedy appointed a committee, headed by Senator Stuart Symington, former Secretary of the Air Force, to study defense organization for the purpose of recommending needed changes. After two months of study the Symington Committee proposed a radical reorganization of DOD.

When Robert McNamara, President of Ford Motor Company, accepted Kennedy's offer to become the eighth Secretary of Defense, he persuaded the young President-elect to hold off proposing the reorganization until he (McNamara) could assess the situation personally. But, Kennedy did give McNamara two directives: first, develop the military structure required for a firm foundation for our foreign policy without regard to budget ceilings; second, operate this force at the lowest possible price.[27]

Armed only with those two directives and without benefit of any new legislation, McNamara made so many changes, both formal and informal, organizational and procedural, that he brought about not just a reorganization but a revolution in DOD. Nowhere was this revolution more acutely felt than in the budget process where the eighth Secretary of Defense introduced the Planning-Programming-Budgeting System (PPBS) and cost effectiveness or systems analysis.[28]

PPBS divided the budgetary process into three clearly defined cycles and lengthened it to 18 months. Thus, preparation of the FY 1966 budget, which was to become effective in July 1965, and was to be submitted to the Congress in January 1965, got underway in July 1963. Cost effectiveness or systems analysis is a technique that looks at alternate ways of performing a job and seeks, by estimating in quantitative terms where possible, to identify the most effective alternative.

The foundation for PPBS in DOD from 1961 through 1968 was the Five Year Defense Plan (FYDP). This document was the master plan for the budget process and contained the programs, approved by OSD with their estimated costs projected for five years. The initial FYDP was produced in 1961 and projected programs and costs through 1965. Each year the FYDP was updated by decisions made during the budget process.

The planning cycle was the first and longest. It began in July and lasted until February and was composed of three steps. The first step involved production of Volume I of the JSOP by the JCS. This volume was an assessment of the military threat facing the United States and of our national commitments projected for five years.

For about five years, the JCS devoted a great deal of time and energy to producing this part of the JSOP. Since there was no BNSP, the Chiefs hoped that their estimates would furnish the basis for all subsequent budgetary decisions. They hoped to make it a substitute for the BNSP.[29] However, by 1965, the JCS realized that the JSOP had little impact on subsequent budget decisions. Thus, they began to spend less and less time on it and by 1967 had turned the job over to their subordinates.[30]

While the JCS were working on the first part of the JSOP, McNamara often assigned special projects to the Chiefs. These projects consisted of a set of specific questions, had very short deadlines, and had great potential implications for the budget. For example, in March 1961, McNamara asked the Chiefs to estimate how many bombers the United States would need in the next decade and set a deadline of six weeks.[31]

The career officials, e.g., JCS, complained a great deal about these short deadlines, but always completed the studies on time. OSD usually found the studies of the career officials lacking in many respects.[32] From 1961 through 1965, OSD usually was content with pointing out the inconsistencies in these studies. But from 1965 onwards, when the position of Assistant Secretary of Defense, Systems Analysis (SA), was established within OSD, SA began to make recommendations of its own and these were usually accepted by the Secretary.[33]

The second step of the planning cycle consisted of the submission of force level recommendations by the services and unified commands to the JCS. These recommendations were based upon the threat and commitments outlined in Volume I of the JSOP.

The third and final step of the planning cycle involved the completion of two major documents. The JCS completed Volume II of the JSOP. This part recommended the optimum force levels necessary to meet U.S. requirements. Although the force levels were supposed to be based on the advice of both the services and unified commanders, the JCS rarely paid attention to the latter's ideas, and Volume II was based primarily upon service inputs. The Chiefs ignored the unified commander's requests because they were unrealistic and because the JCS were wary of losing any power to these men, who were theoretically their equals.[34]

While the JCS were completing the JSOP, OSD produced a Major Program Memorandum (MPM) for each of the mission areas and support activities of the defense budget. These memoranda summarized the OSD position on the major force levels, the rationale for choices among alternatives, and the recommended force levels and funding. Although the MPM were programming documents based upon the planning in the JSOP, their authors in fact ignored the JSOP and the MPM which were both planning and programming documents.[35]

The programming cycle began with the Secretary's receipt of the JSOP and MPM. This cycle lasted about six months, i.e., through the end of August.

McNamara normally reviewed these documents for about 30 days and then provided guidance to the services for preparing Program Change Requests (PCR), i.e., suggested modifications to the FYDP. The primary factor shaping this guidance was the MPM.

The services normally submitted about 300 PCR's annually to the Office of Systems Analysis, whose decisions were nearly always negative. The rejection of the PCR's was attributable to three factors: the services used poor analytical techniques, their requests did not convey any sense of priority in relation to the base program, and most important, their cost.[36]

Theoretically all the program decisions should have been made before budgeting began but this was not the case. Many of the program decisions were negotiated during and after the budgetary cycle.[37] An OSD official reported that in FY 1968 and FY 1969, 90 per cent of the final program decision documents were not written until after December 28, i.e., after the conclusion of the budgetary cycle.[38]

While Systems Analysis was reviewing the PCR's, the JCS were reviewing the MPM. The Chiefs' comments on the memoranda were sent to the Secretary in July. For the remainder of the summer, McNamara and the JCS met about 15 times to discuss the Chiefs' adverse comments.

From 1961 through 1965, the JCS were never united on the major issues raised in the MPM. The Navy objected to the B-70; the Army opposed a 15 carrier fleet; and the Air Force was less than enthusiastic about ABM. In their meetings, McNamara was able to capitalize on these differences and skillfully played one service off against another.[39]

However, from 1966 onward, the JCS worked out their differences prior to meeting with the Secretary and presented a united front to him. For example, the Air Force wanted 35 wings of tactical aircraft and the Navy 17 carriers. Prior to meeting McNamara they agreed on 29 wings and 15 carriers. Similar negotiations were conducted on the ABM.[40]

The Chiefs realized early that McNamara was dividing and conquering but were not able to work out their differences until Taylor stepped down as Chairman and LeMay was pushed out as Air Force Chief of Staff. Taylor was regarded as an administration man and LeMay was an uncompromising crusader for air power. General Wheeler, Taylor's successor as Chairman, refused to bring split opinions to McNamara.[41]

This united front eventually paid off for the JCS. When they were divided, McNamara could carry the day by pointing out the division to the President and Congress. But even McNamara was hesitant about overruling a united or common professional military opinion.[42] Consequently, such items as a nuclear carrier and the ABM, which the Secretary opposed for about 5 years, were eventually approved.

The budgetary cycle officially began in September when the services were asked to prepare their budgets in the traditional categories, i.e., each service separately rather than in program packages, for submission to OSD by October 1. In issuing his call for budget submissions, McNamara emphatically pointed out, year after year, that the services were not to feel bound by any budgetary ceiling, real or imagined. They were to be guided only by decisions made in regard to the MPM and the PCR's. The Secretary repeatedly stated that this country could afford whatever was necessary for defense.[43] Theoretically, the budgetary cycle was to consist only of costing out approved programs.

Despite McNamara's rhetoric, the JCS had a very good idea of what the total and individual service budgets would be.

Sometimes the Comptroller let the service Chiefs know as early as July. On most occasions, it was a simple matter of arithmetic. It was more than a mere coincidence that what this country could afford for defense from FY 1963 through FY 1966, i.e., before the Vietnam buildup, came within one per cent of $46 billion each year and that the Army, Navy and Air Force shares of the budget remained practically the same as under Eisenhower. A service Chief, who served under Eisenhower and McNamara, said that in regard to budget ceilings there was no real difference between either administration.[44] Another Chief remarked, "Weapon systems became more and more difficult to justify as we approached our portion of $46 billion."[45]

Any lingering doubts about a budget ceiling in DOD were shattered during the Vietnam buildup when McNamara directed the services to delete programs that were not urgent, to assume for budgetary purposes that the war would be over by the end of the fiscal year and that during the year there would be no increase in the level of our commitment to Vietnam, and to stretch out maintenance and repair cycles by about 50 per cent.[46]

Actually, the specific figure for defense expenditures, was arrived at in late summer after considerable bargaining, negotiating, and political conflict among the President, the Director of the Bureau of the Budget, the Secretary of the Treasury and

McNamara. This figure was subject to only marginal adjustment in the last four months of the budget process.[47] There is no doubt that the Secretary of Defense had some slight impact upon this figure, but other factors were much more important. In McNamara's first years, Kennedy's campaign promises to develop a strategy of flexible response and to close the missile gap were the determining factors. During the Johnson Administration, the constraining factors were the President's desire to keep his initial budgets under $100 billion and his subsequent determination to wage the War on Poverty and the Vietnam War simultaneously. The Defense Secretary's lack of impact on the size of the defense budget was demonstrated by the four consecutive supplements that he was forced to seek, i.e., from FY 1965 through FY 1969[48] and the strange assumptions under which he forced the military to budget during those same years, e.g., the war in Vietnam would end at the close of the fiscal year.

From 1961 through 1965, the service budget requests exceeded the amount eventually approved by about 10 per cent. However, from 1966 through 1968, as McNamara's standing in the administration waned, the gap between the amount requested and amount granted widened enormously. According to DOD figures, FY 1967 requests were 19 per cent different, FY 1968 over 28 per cent, and FY 1969 over 30 per cent.

From October through December, the Comptroller's office reviewed these budgets. In its review, the office normally initiated some 600 subject issues, i.e., areas of potential savings. Although these issues were theoretically technical, e.g., the cost of a submarine or the cost of equipping an infantry battalion, in fact the issues reflected intuitive feelings on the part of the personnel in the Comptroller's office about where they felt cuts ought to be made.[49] McNamara reviewed the budgets personally and, with the subject issues as a guide, made about 700 budgetary decisions annually. Often his decisions concerned the smallest matters. During his review McNamara consulted with the JCS about 20 times. These consultations took place on the Secretary's terms. McNamara never allowed the Chiefs to set priorities but only asked them to comment on items individually. He was not interested in whether the Chiefs preferred x or y, only their opinion of x. In deliberating about x, the Chiefs never knew if he would ask about y.[50]

In late December, the President met with the Secretary and the JCS for about four hours to discuss the budget. Despite vigorous opposition on the part of many members of the JCS during their "days in court," the President invariably sided with McNamara. Some of the issues raised in these meetings included the B-70, number of Polaris submarines, and pilot shortages.

The Bureau of the Budget was not any more successful than the Chiefs in getting the President to overrule the Secretary. McNamara was able to tell the Congress that the Bureau was not able to change anything in his budget.[51]

NIXON ADMINISTRATION

Melvin Laird, who had observed McNamara's revolution from his seat on the House Subcommittee on Military Appropriations, felt that McNamara's methods had led to over-centralization in decision-making.[52] Accordingly, when he became the tenth Secretary of Defense, Laird instituted certain changes in the defense budget process to redress this situation. The essence of these changes was contained in a "treaty" signed by the Deputy Secretary of Defense, the Service Secretaries and the Chairman of the JCS. This treaty, "negotiated" soon after Laird assumed the helm at the Pentagon, provided that the Secretary of Defense would look to the services and the JCS in the design of forces and that the Systems Analysis Office would limit itself to evaluation and review and not put forward independent proposals of its own. In return for this concession the Secretary of Defense expected the services to work within the ever decreasing budget ceilings.[53]

The length of the process is still about 18 months,[54] the foundation is still the FYDP, and it includes many of the same steps as under McNamara, but as the "treaty" indicates, the emphasis is different[55] and PPB as a system is now non-existent in DOD. The JCS inaugurate the planning cycle by producing Volume I of the JSOP, i.e., the strategic assessment, and sending it to Laird. The Secretary reviews the JSOP and then issues a coordinated, complete and current strategic guidance document for the entire defense community, the Strategic Guidance Memorandum (SGM). This document is essentially the JSOP with some updating and enlargement and is issued in January, e.g., the SGM for FY 1972 was issued in January 1970.

Despite the effort expended in its production, the SGM is basically useless for budgetary purposes. Like the planning documents in other administrations, it is replete with ambiguities and "waffles" the hard questions and thus does little to bring defense programs into accord with Presidential priorities.[56] Because of the poor quality of the SGM, the non-defense members of the Defense Program Review Committee (DPRC)[57] have been anxious to become involved in the production of the SGM. To date Secretary Laird has been successful in keeping them from so doing.[58]

In January, the Secretary also issues a tentative Fiscal

Guidance Memorandum (FGM), projecting dollar constraints for the next five years. While the elements of DOD are reviewing the TFGM, the JCS complete the force structure portion of the JSOP, i.e., Volume II. This is prepared from a purely military perspective, i.e., without regard to the fiscal constraints of the TFGM.

The Secretary reviews the comments on the TFGM and Volume II of the JSOP and then completes the planning cycle by issuing a Fiscal Guidance Memorandum (FGM) in March. The FGM sets definite ceilings on the total budget and on each service. In Laird's three budgets, the FGM has set a figure of from $75 to $70 billion and split the figure evenly among the three services.[59]

The ceilings for the FGM are established within the DPRC. Its constraints have been guided by a desire to roughly balance the budget and an anticipation of how much Congress will allocate for defense.[60] The ceiling for FY 1971 was $75 billion and for FY 1972 it was $70 billion.[61]

The programming cycle begins in April when the JCS draw up a Joint Force Memorandum (JFM), which presents the Chiefs recommendations on force levels and support programs that can be provided within the fiscal constraints of the FGM. The JFM also includes an assessment of the risks in these forces as measured against the strategy and objectives of JSOP, Volume I, and a comparison of the costs of its recommendations with the FYDP. Finally, the JFM highlights the major force issues to be resolved during the year. In 1969 and 1970, these issues have included the B-1, ABM, and shipbuilding.

In May, each service submits to the Secretary of Defense a Program Objective Memorandum (POM) for each major mission area and support activity in the defense budget. These memoranda express total program requirements in terms of forces, manpower, and costs and must provide a rationale for deviations from the FYDP and the JFM. OSD no longer issues program documents.

In July, Laird completes the programming cycle by issuing Program Decision Memoranda (PDM) for each budget area. These are based upon the inputs of the JSOP, JFM, and POM and are then reflected in the FYDP. During July and August, the Secretary meets with the JCS to resolve any disputes over the PDM. These disputes have mainly centered around a manpower-weapons tradeoff. The JCS have opted for decreasing manpower and putting the limited funds into advanced weaponry, e.g., F-14, F-15. In the FY 1971 and 1972 budget evolutions a compromise has been worked out. [62]

The budgetary cycle commences on September 30, when

each service submits its budget to the Secretary. The budgets are supposed to be based on the approved programs resulting from the various decision documents. The service submissions have come within 3 per cent of the established ceiling. In FY 1970 Congress authorized $77.5 billion for defense and in FY 1971 the service requests amounted to $77.3 billion.[63]

After a review of the budget estimates by the OSD staff, working with 51 representatives of the Office of Management and Budget, the budget is sent to the DPRC. This committee reviews the budget in November and December. The DPRC made very few changes to the FY 1971 budget but in December 1970 recommended that an additional $6 billion be added to the FY 1972 budget to maintain troop levels in Europe and the strength of the Sixth Fleet. OSD wanted to reduce both of these items to stay within the $70 billion ceiling. President Nixon ratified the decision when he decided to have a deficit budget.[64]

If the JCS object to any of the decisions made during the process, they are free to take their case outside of DOD. Indeed, Secretary Laird encourages them to state their grievances to individuals like the Secretary of the Treasury, the Director of the Office of Management and Budget, and to bodies such as the DPRC.[65]

CONCLUSION

Perhaps the best way to conclude an analysis of the parts played by the Secretary of Defense and the JCS in the defense budget process is to identify the essential elements in that process. The preceding sections have shown that, despite changes in procedure from administration to administration, there are four steps which are common to all post World War II administrations.[66] They are:

1. Establishment of the overall size of the military budget.

2. Production of strategic plans to guide the distribution of defense funds.

3. Preparation and submission of military requests.

4. Review and revision of these requests.

In regard to the first step, i.e., establishment of the total for defense, this analysis has shown that the Secretary of Defense has had very little impact. President Truman, without consulting his Secretaries, decreed that the military could have one-third of the total federal budget and would not budge from this ceiling even in the face of pleas by Forrestal. Similarly, Eisenhower decided that this country could not afford to spend more than 10 per cent of its GNP for defense. Although his Secretaries agreed with this feeling of the President, the impetus for it came from elsewhere, i.e., most notably the Secretary of the Treasury.

It should be remembered that Secretary Wilson is famous for saying "more bang for the buck," and not "more bucks for banging."

During Robert McNamara's seven year tenure, there were no "magic number" ceilings passed down from the White House. Yet, in spite of McNamara's rhetoric, there were quite specific constraints imposed before the program decisions were made. While McNamara was involved in determining these ceilings to a greater extent than his predecessors, the desires of the President were the primary constraining factors.

Secretary Laird, through his participation in the NSC System, likewise is involved in the bargaining process that determines the defense budget ceiling. However, Laird is one voice among many, and there is no evidence to indicate that he dominates the process. On the contrary, there is evidence that Laird is apparently indifferent to the NSC process and is content to serve as an adjudicator of service interests within the ceilings.[67]

The impact of the JCS upon the amount spent for defense has been less than that of the Secretary. Prior to this administration the Chiefs were not even consulted before the establishment of the ceilings and, in spite of their sometimes[68] strenuous efforts during the budget process, have never raised the ceiling. Under the present administration, the chairman of the JCS is a member of the DPRC, but the fact of declining expenditures since the creation of that body does not appear to indicate that he has had a great impact. General Earl Wheeler, who was a member of the DPRC in 1969 and 1970, has become a supporter of the American Security Council's "Operation Alert," which is attempting to publicize the lack of defense spending.

In every post-World War II administration, the Secretary and the JCS have been involved in the production of plans that were supposed to guide the distribution of the defense dollar. The names of these plans and their relative position in the budgetary process have varied, but all these plans have been almost totally irrelevant to the budgetary process.

In the Truman and Eisenhower Administrations, the Secretary was involved in the planning done by the NSC. But, in light of the stringent budget ceilings during the Truman Administration, NSC planning was totally unrealistic, and the BNSP of Eisenhower's NSC was too vague and ambiguous to have any real impact upon the budget.

Both McNamara and Laird have had their own budgetary plans, the MPM and SGM respectively. However, the majority of the program documents during McNamara's tenure were written after the budgetary decisions had been made, and Laird's SGM suffers from many of the same defects as the BNSP.

The master plan of the JCS during this time period has been known as the JSOP. During the Truman Administration, the JCS were planning to hold the line at the Rhine with a budget that did not even have sufficient funds to maintain a line of communication in the Mediterranean. The JSOP that was produced during the Eisenhower Administration was three separate plans added together. Nobody outside of the JCS read the JSOP while McNamara was holding the reins. The current JSOP is the foundation for the SGM.

The main beneficiaries of this gap between planning and budgeting have been the service chiefs. With no operative framework, the Chiefs have been free to request nearly anything that they want in the third essential step, i.e., preparation and submission of the monetary requests. This analysis has demonstrated that this third step has been the sole perogative of the individual service chiefs all throughout this period. Neither the reorganizations of DOD,[69] the creation of unified commands,[70] nor the institution of program packages has weakened JCS control over the preparation and submission of the separate service budgets.

Moreover, because of the absence of any real link between the plans and the budgets, such factors as program decisions, and administration priorities seem to have had little impact upon the distribution of funds in these requests.[71]

The Secretary of Defense's greatest impact upon the budgetary process has been in the area of reviewing and revising the service requests, i.e., the fourth essential step. All of the secretaries have made substantial revisions but none has made as many or been as free from further alterations as Robert McNamara. Forrestal and Johnson made many modifications to the service budgets, but their DOD budgets were significantly changed by the Bureau of the Budget. Similarly, many of the budgetary decisions of Wilson and McElroy were altered by Eisenhower. However, McNamara, who made some 700 detailed budgetary decisions annually, could boast to Congress that not one single thing was ever changed in his budget once it left DOD. Because of the closeness of the JCS submissions to the ceilings, Secretary Laird does not have too many revisions to make, and those that he does are subject to scrutiny by the DPRC.

The JCS, on the other hand, have been almost totally without influence during the budgetary review. During the administrations of Truman and Eisenhower, they could have dominated that phase but chose not to participate. Under McNamara, although they were consulted by the Secretary during the review, the Chiefs were not able to establish priorities. During the present administration, the review process has not been as significant because the service requests have very nearly coincided with the ceilings.

Looking at the four essential steps of the budgetary process, it is clear that the total for defense spending is basically decided outside DOD, but that the distribution of funds within that ceiling has been basically controlled by the Secretary and the JCS. Moreover, this distribution has not been related to a coordinated plan in any meaningful way and has, for the most part, been dominated by the JCS. Much of the Secretaries' reviews have been of the across the board variety, e.g., 10 per cent in manpower, and their total reductions rarely exceed 15 per cent.[72] Furthermore, when a Secretary seeks to become more deeply involved, e.g., Johnson or McNamara, the Chiefs have blunted his impact by submitting requests that are very close to the ceilings or by compromising their differences and presenting a united front on the important budget issues.

In looking to the future of the size and distribution of the military budget two things seem obvious. First, the level of defense expenditures, expressed as a part of the total budget and as a percentage of the GNP, will continue to go down. Defense expenditures for FY 1972 will be below 30 per cent of the total federal budget and below 7 per cent of the GNP.[73]

The cause of this relative decline in defense expenditures

is the attitude of Congress and the public, and not the desires or role perceptions of the leadership of DOD or of the administration.[74] This administration, as others before it, anticipates the reaction of the Congress and the public in setting its ceiling on defense expenditures and is well aware that Congress will not appropriate more than about $70 billion. It should be borne in mind by those who feel that this nation has spent too much on defense in the past that, prior to Vietnam, Congress and the public favored spending in excess of the administration's ceilings.[75]

Second, the distribution process will continue to be dominated by the top military professionals, i.e., the JCS.[76] Unless the Secretary of Defense is able to have his own military staff, e.g., the Joint Staff, he will have to continue to negotiate treaties with the JCS. The prospects for this do not seem bright. DOD was reorganized in 1949, 1953, and 1958, and revolutionized in 1961 without really changing the service bias of the Joint Staff. To date none of the recommendations of the Fitzhugh Panel along this line have been implemented.

Doubtlessly, some secretaries will have more impact on the budget than others, primarily because of their role perceptions. But, it should be remembered that Robert McNamara, who had the broadest possible conception of his position and who was the most forceful, and dynamic Secretary of Defense in history, could not overrule unanimous JCS opinions on such important issues as the ABM and the nuclear powered carriers.

NOTES

This paper was originally presented at the Sixth Annual Conference of the Inter-University Seminar on Armed Forces and Society, November 18-20, 1971, Chicago, Illinois.

Because many of the footnotes in this paper will come from the annual congressional hearings on the defense budget, a simplified system to reduce the citations to manageable proportions has been used. References to the hearings of the committees will be found in the following form: HCA or HCAS, 1965, I, 57 and SCA, 1965, II, 95. HCA and SCA refer to the House and Senate Appropriations Committees before which the hearings on the defense budget are conducted, the year refers to the fiscal year for which the money will be appropriated, and the Roman numerals signify the volume number.

[1] In referring to the JCS and the Secretary we mean not only the individuals but their staffs, i.e., the Organization of the JCS (OJCS) and the Office of the Secretary of Defense (OSD).

[2] Few men are able to step into a job as complex as Secretary of Defense and make major changes immediately. Even the dynamic Robert McNamara submitted a budget prepared by the Eisenhower Administration to the Congress.

[3] Harry Truman, Years of Trial and Hope (Garden City: Doubleday, 1956), p. 37. And Samuel Huntington, The Common Defense (New York: Columbia University Press, 1961), p. 42.

[4] JCS Plans envisioned "Holding the Line at the Rhine." SCA, 1950, p. 17. The eventual budget allowed them only to mount an air offensive from Great Britain.

[5] Walter Millis, ed. The Forrestal Diaries (Viking Press: New York, 1951), p. 415. Work on the basic FY 1949 budget was completed before Forrestal took office.

[6] SCA, 1950, 17.

[7] Forrestal Diaries, pp. 418 and 500.

[8] Ibid. These intermediate budgets were predicated upon maintaining a Line of Communication (LOC) in the Mediterranean.

[9] Ibid., p. 430 and SCA, 1950, 17.

[10] SCA, 1950, 32.

[11] For a complete account of the decision see Paul Hammond, "Supercarriers and B-36 Bombers," in *American Civil-Military Decisions*, (Tuscaloosa: University of Alabama Press, 1963).

[12] The events of the executive phase of the FY 1951 budget were extracted from the testimony of Wilfred McNeil, DOD Comptroller, HCA, 1951, I, 104-06; and budget Director Frank Pace, HCA, 1951, I, 2659-2664, and an interview with McNeil.

[13] The format for Eisenhower's budget was obtained from interviews with McNeil, Burke, Taylor, and Radford, and Maxwell Taylor, *The Uncertain Trumpet* (New York: Harper, 1959) and Matthew Ridgway, *Soldier* (New York: Harper, 1956).

[14] Taylor, *Trumpet*, pp. 82-83.

[15] Alain Enthoven, and K. Wayne Smith, *How Much is Enough* (New York: Harper and Row, 1971), pp. 12-13.

[16] Interview: Secretary of the Treasury, George Humphrey, was the leading spokesman for the idea of a balanced budget. In January 1957, he told a news conference that there were many places where Eisenhower's FY 1958 budget could be cut by Congress. HCA, 1958, 5. Congress responded by lopping $2.5 billion off of a $38 billion budget.

[17] Congress was dismayed when Eisenhower's FY 1959 budget, submitted about three months after Sputnik, amounted to only $37 billion. HCA, 1959, 353.

[18] HCA, 1964, II, 585.

[19] These men felt it was their job to increase savings and efficiency in the Pentagon.

[20] HCA, 1954, 317, 470-71.

[21] HCA, 1955, 43-45. Wilson and Radford tried to convince the JCS that a sound economy was an aspect of national security and a proper subject for military appraisal.

[22] Nathan Twining, *Neither Liberty Nor Safety* (New York: Holt, Rhinehart and Winston, 1966), p. 56.

[23] Taylor, *Trumpet*, p. 78.

[24] Senate Preparedness Investigating Subcommittee, *Hearings on Major Defense Matters* (Washington, D.C.: U.S. Government Printing Office, May 20, 1959), p. 206.

[25] In a perceptive article, Donald Gumz, Captain, USN, "The Bureau of the Budget and Defense Fiscal Policy," *U.S. Naval Institute Proceedings* (April, 1959), pointed out that the fact that such detailed decisions were made by civilians was the fault of the JCS. McNeil did consult with the JCS on an individual basis.

[26] Interview. The practice of the Bureau of the Budget appealing to the President is the reverse of other departments. It still continues.

[27] HCA, 1963, 4.

[28] The best source on McNamara's methods is Enthoven and Smith, *How Much is Enough*. The format for McNamara's budget procedures is adopted primarily from this book. I also relied on interviews with Taylor, Decker, Anderson, McDonald, McConnell, and Barber. John Crecine, *Defense Budgeting: Organizational Adaptation to External Constraints*, Rand, March, 1970, and William Kaufmann, *The McNamara Strategy* (New York: Harper, 1964), are also excellent sources.

[29] Interview

[30] Interviews

[31] HCA, 1963, II, 4-6.

[32] Enthoven and Smith, p. 32.

[33] Interview

[34] Interview

[35] Enthoven and Smith, p. 94, call the JSOP the "best example of the unrealistic alternatives provided by the military." Enthoven's office wrote the MPM.

[36] HCA, 1968, I, 88. William Niskanen, "The Defense Resources Allocation Process," *Defense Management*, (edited by Stephen Enke), (Englewood Cliffs: Prentice Hall, 1967), p. 10. The PCR's would have added about $40 billion annually.

[37] Enthoven and Smith, p. 56.

[38] Interview: OSD official, April, 1968, quoted in Crecine, p. 41.

[39] Interview

[40] Ibid.

[41] Ibid. One interviewee said, "LeMay had his head in concrete."

[42] Ibid.

[43] HCA, 1963, II, 4-6. HCA, 1965, I, 304.

[44] Interview

[45] Ibid.

[46] HCA, 1967, I, 280; SCA, 1967, I, 69.

[47] John Crecine and Gregory Fischer, On The Resource Allocation Processes in the U.S. Department of Defense (Institute of Policy Studies Discussion Paper No. 31, The University of Michigan, October, 1971), pp. 13-14 present an excellent discussion of the process by which the ceiling was established.

[48] Phillip Goulding, who was McNamara's assistant for Public Affairs, feels that these supplements undermined the Secretary's credibility with the President and contributed to his dismissal. Confirm or Deny, (New York: Harper and Row), pp. 168-214.

[49] April 1968 Interview with an official in the Comptroller's Office, quoted in Crecine, p. 51.

[50] Interview. HCA, 1965, IV, pp. 447-95.

[51] SCA, 1964, 211-212.

[52] HCA, 1971, I, 153.

[53] Enthoven and Smith, p. 334. Interviews. DOD is caught in a crossfire between public and congressional sentiment demanding less spending on defense and inflation pushing costs upward.

[54] Laird's budget procedures are outlined in HCA, 1971, III, 480-81.

[55] Crecine and Fischer, p. 35.

[56] Interview: Defense Program Review Committee (DPRC) Working Group Member, March 1971. These ambiguities are also

reflected in the <u>Defense Report</u>, an unclassified version of the SGM, which is presented to Congress annually by the Secretary of Defense.

[57] The DPRC is a subcommittee within the NSC system composed of the President's Assistant for National Security Affairs, the Deputy Secretary of Defense, the Undersecretary of State, the Chairman of the JCS, the Director of the Office of Management and Budget, and the Chairman of the President's Council of Economic Advisors. This body was created in October 1969 to anticipate the political, economic and social implications resulting from changes in defense spending, budgeting and force levels.

[58] Interview: DPRC Working Group Member.

[59] In Nixon's three budgets the Army has received $67.7 billion, the Navy $67.8 billion and the Air Force $69.9 billion.

[60] When Congress reduced the FY 1970 budget by $5.7 billion, the ceiling for FY 1971 was reduced by a similar amount. HCA, 1971, III, 480-81.

[61] Interview: DPRC Working Group Member.

[62] <u>Ibid</u>. Up to now, Laird has been a shrewd adjudicator of service interests.

[63] HCA, 1971, I, 59.

[64] Interview: DPRC Working Group Member.

[65] Secretary Laird: Press Conference August 14, 1971.

[66] I have argued elsewhere that the essentials of the budgetary process have remained unchanged since the creation of DOD. "The Defense Budget Process in the United States, 1953-1970: An Examination and An Evaluation," presented at the 1971 Meeting of the American Political Science Association, Chicago. Crecine and Fischer make a similar point.

[67] Interviews. See also John Leacacos, "Kissinger's Apparat," <u>Foreign Policy</u> (Winter 1971-72), p. 4, and I. M. Destler, "Can One Man Do," <u>Foreign Policy</u> (Winter 1971-72), p. 34.

[68] There have been occasions when the Chiefs have accepted the ceilings without question, e.g., FY 1951.

[69] One of the main purposes of the 1949, 1953 and 1958 Reorganizations has been to weaken the power of the JCS.

[70] The JCS usually defer to the unified commanders, e.g., CINCPAC, on operational matters. Morton Halperin, "The President and the Military," Foreign Affairs (January 1972), p. 321.

[71] Crecine and Fischer, p. 5, have shown that Kennedy's $8 billion dollar addition, which was allegedly for the purpose of closing the missile gap and increasing the readiness of conventional forces, resulted in very little percentage reallocation among either the services or appropriation categories.

[72] HCA, 1964, II, 585.

[73] Former budget director Charles Schultze estimates the defense spending will be down to 26 percent by FY 1976. Charles Schultze et. al., Setting National Priorities (Washington: The Brookings Institution, 1971), p. 325.

[74] William Baroody, Special Assistant to Laird, asserted that congressional cuts "have impacted directly on our capabilities." Navy Times (January 19, 1972), p. 3.

[75] From FY 1959 through FY 1963 Congress appropriated over $2 billion more than the administration requested for defense and tried a variety of devices to compel the administration to spend the additional funds.

[76] The Chiefs are already competing innovatively to adopt their traditional arms to the aseptic connotations of the Nixon Doctrine, Earl Ravenal, "The Political-Military Gap," Foreign Policy (Summer 1971), p. 39.

Interviewees

Anderson, George, Admiral USN, CNO, September 5, 1968.

Barber, Art, Deputy Assistant Secretary of Defense, February 2, 1968.

Burke, Arleigh, Admiral USN, CNO, September 6, 1968.

Decker, George, General USA, Chief of Staff, Sept. 4, 1968.

McConnell, John, General USAF, Chief of Staff, June 17, 1971.

McDonald, David, Admiral USN, CNO, August 27, 1968.

McNeil, Wilfred, Comptroller, December 13, 1968.

Nutter, G. Warren, Asst. Sec. of Defense, January 29, 1971.

Radford, Arthur, Admiral USN, Chairman JCS, September 4, 1968.

Sikes, Robert, Chairman of the Subcommittee on Military Construction, August 14, 1968.

Spaatz, Carl, General USAF, Chief of Staff, September 5, 1968.

Taylor, Maxwell, General USA, Chairman JCS, Chief of Staff, Military Assistant to the President, September 4, 1968.

Wheeler, Earl, General USA, Chairman JCS, Chief of Staff, October 8, 1971.

UA
23
.M562

The Military-
industrial complex:
a reassessment

31049

Date Due

JAN 24 '74			
OCT 16 1975			
DEC 4 1975			
APR 19 1978			
MAY 3 1979			
MAY 21 1981			
NOV 19 1987			
APR 2 1999			
MAY 28 2010			

The Library

COLLEGE of MARIN

Kentfield, California

 PRINTED IN U.S.A.